DISCOVERING
the
EARTH

PLANTS

Food, Medicine, and the Green Earth

DISCOVERING
the
EARTH

PLANTS

Food, Medicine, and the Green Earth

Michael Allaby
Illustrations by Richard Garratt

Facts On File
An imprint of Infobase Publishing

PLANTS: Food, Medicine, and the Green Earth

Facts On File, Inc.
An imprint of Infobase Publishing
132 West 31st Street
New York NY 10001

Library of Congress Cataloging-in-Publication Data

Allaby, Michael.
  Plants : food, medicine, and the green earth / Michael Allaby ; illustrations by Richard Garratt.
      p. cm. — (Discovering the Earth)
  Includes bibliographical references and index.
  ISBN 978-0-8160-6102-0
  1. Plants—History. 2. Botany—History. I. Facts on File, Inc. II. Title. III. Series: Allaby, Michael.
Discovering the Earth.
  QK45.2.A45 2010
  580—dc22   2009018930

Facts On File books are available at special discounts when purchased in bulk quantities for businesses, associations, institutions, or sales promotions. Please call our Special Sales Department in New York at (212) 967-8800 or (800) 322-8755.

You can find Facts On File on the World Wide Web at http://www.factsonfile.com

Text design by Annie O'Donnell
Composition by Hermitage Publishing Services
Illustrations by Richard Garratt
Photo research by Tobi Zausner, Ph.D.
Cover printed by Times Offset (M) Sdn Bhd, Shah Alam, Selangor
Book printed and bound by Times Offset (M) Sdn Bhd, Shah Alam, Selangor
Date printed: March, 2010
Printed in Malaysia

CONTENTS

PREFACE

Almost every day there are new stories about threats to the natural environment or actual damage to it, or about measures that have been taken to protect it. The news is not always bad. Areas of land are set aside for wildlife. New forests are planted. Steps are taken to reduce the pollution of air and water.

Behind all of these news stories are the scientists working to understand more about the natural world and through that understanding to protect it from avoidable harm. The scientists include botanists, zoologists, ecologists, geologists, volcanologists, seismologists, geomorphologists, meteorologists, climatologists, oceanographers, and many more. In their different ways all of them are environmental scientists.

The work of environmental scientists informs policy as well as providing news stories. There are bodies of local, national, and international legislation aimed at protecting the environment and agencies charged with developing and implementing that legislation. Environmental laws and regulations cover every activity that might affect the environment. Consequently every company and every citizen needs to be aware of those rules that affect them.

There are very many books about the environment, environmental protection, and environmental science. Discovering the Earth is different—it is a multivolume set for high school students that tells the stories of how scientists arrived at their present level of understanding. In doing so, this set provides a background, a historical context, to the news reports. Inevitably the stories that the books tell are incomplete. It would be impossible to trace all of the events in the history of each branch of the environmental sciences and recount the lives of all the individual scientists who contributed to them. Instead the books provide a series of snapshots in the form of brief accounts of particular discoveries and of the people who made them. These stories explain the problem that had to be solved, the way it was approached, and, in some cases, the dead ends into which scientists were drawn.

There are seven books in the set that deal with the following topics:

- Earth sciences,
- atmosphere,
- oceans,
- ecology,
- animals,
- plants, and
- exploration.

These topics will be of interest to students of environmental studies, ecology, biology, geography, and geology. Students of the humanities may also enjoy them for the light they shed on the way the scientific aspect of Western culture has developed. The language is not technical, and the text demands no mathematical knowledge. Sidebars are used where necessary to explain a particular concept without interrupting the story. The books are suitable for all high school ages and above, and for people of all ages, students or not, who are interested in how scientists acquired their knowledge of the world about us—how they discovered the Earth.

Research scientists explore the unknown, so their work is like a voyage of discovery, an adventure with an uncertain outcome. The curiosity that drives scientists, the yearning for answers, for explanations of the world about us, is part of what we are. It is what makes us human.

This set will enrich the studies of the high school students for whom the books have been written. The Discovering the Earth series will help science students understand where and when ideas originate in ways that will add depth to their work, and for humanities students it will illuminate certain corners of history and culture they might otherwise overlook. These are worthy objectives, and the books have yet another: They aim to tell entertaining stories about real people and events.

—Michael Allaby
www.michaelallaby.com

ACKNOWLEDGMENTS

All of the diagrams and maps in the Discovering the Earth books were drawn by my colleague and friend Richard Garratt. As always, Richard has transformed my very rough sketches into finished artwork of the highest quality, and I am very grateful to him.

When I first planned these books, I prepared for each of them a "shopping list" of photographs I thought would illustrate them. Those lists were passed to another colleague and friend Tobi Zausner who found exactly the pictures I felt the books needed. Her hard work, enthusiasm, and understanding of what I was trying to do have enlivened and greatly improved all of the books. Again I am deeply grateful.

Finally, I wish to thank my friends at Facts On File, who have read my text carefully and helped me improve it. I am especially grateful for the patience, good humor, and encouragement of my editor, Frank K. Darmstadt, who unfailingly conceals his exasperation when I am late, laughs at my jokes, and barely flinches when I announce I'm off on vacation. At the very start, Frank agreed this set of books would be useful. Without him they would not exist at all.

INTRODUCTION

Life would be impossible without plants. They provide our food, either directly or in the form of plant-eating animals. They supply many of the fibers we use, the timber to build our homes, our fuel for heating and cooking, and wood to make tools and furniture. It is not surprising, therefore, that people have been studying plants for thousands of years. This book tells of that search for understanding and of some of the individuals who contributed to it.

Plants, one volume in the Discovering the Earth set, begins the story in ancient Greece, where philosophers first speculated about the origin of plants, and the study of botany began. The book then considers medicinal plants, which were of prime concern to the founders of botany. Until modern times, most medicines were obtained from plants and physicians were taught botany so they could identify herbs with therapeutic properties. Then, rather than spending time collecting medicinal herbs in the wild, people began to cultivate them in gardens, a practice that may have begun in China, but that flourished in Europe. At the same time as they were cultivating useful plants, botanists were also listing them in books that described each plant, with instructions for its cultivation and uses.

In time, the herbal—a list of medicinal and culinary herbs—grew into the flora—a list of plants of all kinds—and gardens designed on the lines of monastery herb and vegetable gardens expanded to become botanical gardens, exhibiting a wide range of plants for mainly educational purposes. The development of the concept of the botanical garden was soon linked to botanical explorations overseas, and botanists were dispatched to far regions of the world to collect specimens of exotic plants that could be grown back home. Many of the plants they brought to the botanical gardens later found their way into private gardens, where they were grown for ornament. *Plants* recounts the adventures of some of the most famous plant collectors.

As botanists recorded more and more of the plants growing in their own countries and as increasing numbers of plants arrived from abroad, the ever-lengthening lists of plant names became very unwieldy. Botanists—and gardeners—needed a simple and unambiguous system for classifying plants that would allot a unique name to each type. The book tells of the way such a system developed, finally becoming the classification that scientists use today.

Overseas exploration also led botanists to consider the way plant species are distributed globally. They speculated about why particular plants grew where they did, and they observed that similar types of vegetation occurred in regions with similar environmental conditions, but that those vegetation types were often composed of entirely different and only distantly related species. The study of plant geography led to speculation about the ways plants adapt to their environment, which led in turn to the investigation of plant evolution.

Thousands of years ago our ancestors obtained their plant foods— fruits, seeds, leaves, and roots—by gathering them from plants growing wild. Then they began to cultivate some of those plants and very slowly the plants were transformed into the domesticated crop plants that farmers grow today. The book describes the origin of some of the most important crops and tells a few of the stories associated with them. These include a few colorful legends, but also historical events that have given us chocolate and rubber, as well as the famous mutiny on board HMS *Bounty*.

Plants do not grow in isolation, and by the 19th century botanists were beginning to study them as communities. This branch of botany developed into plant sociology, which was also a branch of plant ecology. The book describes the early development of ecology. *Plants* ends with an explanation of biodiversity and of why scientists think it important.

People have been thinking about plants and studying them for thousands of years. In the course of that long history they have accumulated vast amounts of information, and even if it were possible to compress all of that into a single volume the result would be unwieldy and confusing. This book makes no such pretense. It amounts to nothing more than a series of snapshots providing brief glimpses of some of the events that have led to the present scientific understanding of plants and short accounts of the lives of a few of the remarkable individuals who have contributed to that understanding. The book has a glossary defining the technical terms used in the text, and for readers who would like to pursue the subject further there is a list of books and Web sites where they will be able to learn more.

The Father of Botany

Natural history is the study of plants and animals, with the emphasis on observation rather than experiment. People have always made use of plants, of course, but as means to an end, a resource to be utilized. They gathered plants for food and medicines and later they cultivated them; they constructed their buildings, boats, and many tools and utensils from plant materials; and they wove cotton, linen, hemp, and jute to make cloth. Once writing was invented, people made paper from papyrus (a type of sedge), hemp, and cotton, before turning to wood pulp in modern times; all of these are plant products. Communities had been doing these things for thousands upon thousands of years, but people gathering or cultivating plants and craftspersons converting plant materials into useful articles had no need to understand any more about the plants they used than their properties and how to exploit them.

It was not until the cultural flowering in ancient Greece that scholars began to speculate about the origins of plants, to investigate their growth and structure, and to attempt to catalog the many different kinds of plants. That was the beginning of natural history, and in later centuries it led to the study of botany and eventually to the modern scientific disciplines that are grouped together as the life sciences.

This chapter tells of the beginnings of natural history. The story starts in Greece, moves to Rome, and finally arrives in England, where it describes one of the first serious attempts to list every type of plant in the world.

ARISTOTLE AND HIS NATURAL HISTORY

Aristotle (384–322 B.C.E.) was a Greek philosopher who lived at a time when most people believed in a world governed by gods, demigods, and other supernatural beings. According to tradition, in the distant past these beings had made the plants and the stories about how they did so are woven into the Greek myths. It was Gaia, the Earth, for instance, who made the apple tree, as a wedding gift for the goddess Hera, who married Zeus. When the goddess Demeter was searching for her daughter, Persephone, a farmer called Phytalos welcomed her into his home. The goddess rewarded him by creating the fig tree and placing it on his land.

There was a traditional explanation for almost everything, and Aristotle rejected all of them. One of the most famous teachers the world has ever known, he taught his students that they should never accept anything as being true simply because tradition or a person in authority said it was. The only way they could acquire true knowledge of the natural world was by observing it.

Aristotle based his ideas about the natural world on a philosophical system. He believed that all matter had the potential to become form. For example, a seed was matter with the potential to grow into a plant, which was the form; an embryo was matter with the potential to grow into the form of an animal; and a block of stone was matter with the potential to be fashioned into the form of a sculpture. Everything in the natural world had a function and lay somewhere on a scale between pure matter and pure form, with matter lacking form at one end of the scale and form without matter at the other end. The scale was one of values, because as matter progressively acquired form its degree of organization increased. Among living things, Aristotle placed plants at the bottom, animals above plants, and humans above animals.

In his scheme every living thing possessed a soul, but by soul Aristotle meant the completed form of the original matter. In a sense it was the truth contained in the matter that had been revealed in its form. Plants had a soul containing a nutritive element that allowed them to grow and reproduce. Animal souls contained an appetitive element that allowed them to have sensations and desires and in order to satisfy those desires the appetitive element gave them the ability to move. Human souls had both the nutritive and appetitive features, but also a rational element, which gave them the capacity for thought.

Aristotle studied plants and animals and certainly dissected some animals. He wrote only about animals, however, leaving it to Theophrastus (ca. 371–ca. 287 B.C.E.; see "Theophrastus, the Father of Botany" on pages 4–6), his student and successor, to write about plants.

Aristotle was born in 384 B.C.E. at Stagirus, a Greek colony on the coast of Macedon (modern Macedonia). Both his parents were Greek and his father, Nichomachus, was the personal physician to Amyntas III, the king of Macedon. Nichomachus died when Aristotle was still a child and a guardian Proxenus raised him. When Aristotle was 17, Proxenus sent him to Athens to study at the Academy led by Plato (428 or 427–348 or 347 B.C.E.). Aristotle left Athens after the death of Plato, settling first in Anatolia in what is now Turkey and later on the island of Lesbos, where he lived from 345 to 343 B.C.E. The following map shows the geography of the region at that time.

It was while on Lesbos that Aristotle spent much time studying marine animals. In 343 B.C.E., Aristotle returned to Macedon, where

The eastern Mediterranean region as it was in the time of Aristotle. His birthplace, Stagirus, was on the three-pronged peninsula on the coast of Macedon (Macedonia). The various city-states were often at war, but people from all the states visited festival sites. Anatolia was the part of the Persian Empire to the north of Lesbos.

Aristotle (384–322 B.C.E.) in his later years. The photograph is of a Roman marble bust that is a copy of a Greek original, which is now lost. *(The Granger Collection)*

King Amyntas had died and his son had succeeded him as Philip II. Philip appointed Aristotle as tutor to his 13-year-old son, Alexander (see "Alexander the Great and His Empire" on pages 10–13).

When Alexander became king he had no more time for lessons, and in about 335 B.C.E. Aristotle returned to Athens, where for the next 12 years he taught at the Lyceum, one of the three most famous schools in the city. The school was located in the grounds of the temple to Apollo Lyceius, hence its name. The following illustration shows him as he may have appeared at around this time. There were colonnades—covered walkways—at the school, and Aristotle liked to walk through them, surrounded by students, while lecturing. The Greek word for colonnades is *peripatoi*, and the Lyceum school came to be known as the Peripatetic school. Alexander died in 323 B.C.E., and anti-Macedonian feelings began to run high in Athens. Aristotle was charged with impiety, a crime that could have carried a death penalty. Rather than stand trial, he left Athens and settled in Chalcis (see the map on the previous page). He died the following year.

THEOPHRASTUS, THE FATHER OF BOTANY

When Aristotle quit Athens for the last time, he left his friend and assistant Theophrastus (ca. 372–ca. 287 B.C.E.) to lead the Peripatetic school. In his will Aristotle bequeathed the Lyceum buildings and garden and his library to Theophrastus and made Theophrastus guardian of his children. Theophrastus was a popular teacher, some accounts claiming he had 2,000 students, and he lived to a good age. His dying words are alleged to have been a complaint that he was just beginning to gain an insight into life's problems.

Theophrastus was not his real name. He was born as Tyrtamus, and Theophrastus, which means "divine speech," was a nickname, probably given to him by Aristotle and referring to his skills with the spoken word. All that is known about the life of Theophrastus comes from *Lives and Opinions of Eminent Philosophers,* a book by Diogenes Laërtius, who lived about 400 years later and about whom even less is known, not even the years he was born and died. According to Diogenes Laërtius, Theophrastus wrote 227 major works as well as a number of shorter ones. Most of these have survived only as titles or fragments, and they consist of what appear to be lecture notes rather than the texts of books. The two exceptions are Theophrastus's most important works: *De historia plantarum* (On the history of plants) and *De causis plantarum* (On the reasons for plant growth). With these works Theophrastus initiated the methodical study of plants.

Theophrastus grew plants in his own botanical garden and he encouraged his students, many of whom lived a long way from Athens, to observe the plants that grew near their homes. Theophrastus described more than 500 plant species and varieties, classifying them as trees, shrubs, undershrubs, and herbs. He was the first person to distinguish between *monocotyledons* and *dicotyledons*. Monocotyledons produce a single seed leaf (*cotyledon*) and the leaves have parallel veins (grasses are typical); dicotyledons produce two or more cotyledons and the leaves have a network of veins (cabbages are typical). He recognized a fundamental difference between trees that produce cones, such as pines and firs, and those that bear true flowers, such as oaks and aspens. He recorded the different ways plants can reproduce—from seed, cuttings, or roots—and noted that when cultivated trees were grown from seed they often reverted to the wild type, but wild trees did not change from one generation to the next. He described seed germination and the anatomy of different types of flowers, noting that some flowers have petals and others have none.

Not surprisingly, for centuries *De historia plantarum* and *De causis plantarum* were the basic texts for teaching botany. They were translated into Latin in 1483 and again in 1497 and a German translation appeared in 1822. Theophrastus's reports were usually accurate, although he relied for his information about African and Asian plants on accounts by individuals who had taken part in the campaigns of Alexander the Great (see "Alexander the Great and His Empire" on pages 10–13).

De historia plantarum consisted of nine volumes. Volume 1 described plant anatomy. Volumes 2 through 5 were on woody plants, including instructions on cultivation, the treatment of diseases, and the uses and treatment of wood. Volume 6 described herbaceous perennials, volume 7 vegetables and their cultivation, volume 8 cereals, peas, and beans, and volume 9 saps and medicines derived from plants. *De causis plantarum* consisted of six volumes. Volume 1 was on plant reproduction and growth, volume 2 on the environmental factors that affect plants, volume 3 on plant cultivation, volume 4 on the origin and propagation of cereals, volume 5 on plant diseases, and volume 6 on plant flavors and odors. Theophrastus richly deserved the title of father of botany.

Theophrastus was born in about 371 at Eresus (modern Eressos) on the Greek island of Lesbos. He commenced his education on Lesbos, where his teacher Leucippus (or Alcippus) introduced him to the philosophy of Plato. Theophrastus enrolled at Plato's Academy in Athens, and when Plato died he became a follower of Aristotle, probably remaining with him during the time Aristotle spent in Macedon. By all accounts Theophrastus was a kind, generous man and highly popular. When an attempt was made to bring a charge of impiety against him, the case collapsed. After his death in about 287 B.C.E. Theophrastus was given a public funeral, and Diogenes Laërtius wrote that the entire population of Athens turned out to honor him.

MEDICINE AND PLANTS

In Uganda's Kibale National Park, scientists have observed chimpanzees (*Pan troglodytes*) searching for and then chewing the bark or leaves of plants that have very little nutritive value, but that local people use to relieve symptoms of malaria and diarrhea. The chimps also chew these plants when they are sick and they, too, use them in order to rid themselves of intestinal worms as well as to treat malaria and diarrhea. White-faced capuchin monkeys (*Cebus capucinus*) in Costa Rica rub plant material on their bodies, using plants that are known to have insect-repellent and other medicinal properties.

There are many instances of nonhumans—and not only primates—self-medicating with plant substances, and it seems obvious that humans must have been doing so since long before history came to be written down. It is most likely that in prehistory most human communities included a healer who relied on herbal preparations. As in many modern communities, the necessary skills would have been passed from generation to generation, but healers would also have studied the behavior of nonhuman animals in order to learn which plants to use, and almost certainly that is how the tradition began.

Depictions of medicinal herbs in the cave paintings at Lascaux, France, that are between 13,000 and 25,000 years old are the earliest record of herbal medical treatments. There is also physical evidence from antiquity. In 1991 melting of the Similaun Glacier in the Ötzal Alps, Austria, exposed the mummified body of a man, subsequently nicknamed Ötzi. Ötzi died 5,300 years ago, aged about 45. He was carrying with him two pieces of dried birch fungus (*Piptoporus betulinus*) about 1.5 inches (4 cm) in diameter, each of them pierced, and both of them threaded on a single leather thong, perhaps so they could be attached to his belt. Birch fungus, also called razor strop and birch bracket, is a bracket fungus common on birch trees (*Betula* species). Dried, it can be used as tinder, and its cut surface was formerly used to finish sharpening very keen blades such as razors. When scientists first came across Ötzi's dried birch fungus they assumed he carried it as tinder, but later they changed their minds. Birch fungus possesses antibiotic and *antihelminthic*—expelling parasitic worms—properties, and he might have been taking it to combat intestinal parasites.

The earliest written records include lists and descriptions of the uses of medicinal herbs. The Sumerians, living in part of what is now Iraq, were using medicines derived from plants 5,000 years ago, and in about 2000 B.C.E. King Assurbanipal of Sumeria commissioned the first catalog of medicinal plants, describing about 250. Plants were being used in this way 3,000 years ago in ancient Egypt, in India more than 2,500 years ago, and nearly 3,000 years ago in China (see "Shennong, the Divine Farmer" on pages 20–22). Indeed, until the rise of the modern pharmaceutical industry and the associated demand that therapies be administered only by licensed practitioners, herbal remedies were the mainstay of medical treatment, as they are still in many parts of the world.

PEDANIUS DIOSCORIDES AND HIS CATALOG OF MEDICINAL PLANTS

So many plants possess medicinal properties, and combinations of ingredients from different plants generate yet more treatments, that a specialist worker preparing medicines or a physician prescribing them could not possibly remember them all. Medical workers need reference books listing recipes for medicines, descriptions of useful plants and their properties, and aids to diagnosis. For almost 1,500 years until the end of the 15th century and the appearance of the first official pharmacopoeias (see the sidebar that follows), one book was the standard pharmaceutical text. Translated into many languages and published in many editions, it described approximately 600 plants and plant products as well as a few mineral and animal products, 1,000 medicines made from them, and close to 5,000 therapeutic uses for those medicines.

The book was written in about the year 77 C.E. by a Greek physician Pedanius Dioscorides (ca. 40–ca. 90 C.E.), was entitled *De materia medica* (About medicinal substances), and was in five volumes. The first volume described oils, gums, and other aromatic substances. Volume 2 was about animal products including milk and honey, as well as fats, cereals, and herbs. Volumes 3 and 4 were about roots and other herbs, and volume 5 dealt with vines and wines and also mineral preparations using mercury, arsenic sulfide, lead acetate, copper oxide, and calcium hydrate. Dioscorides was less interested in botanical descriptions than in the medicinal uses of plants, and he supplied no more descriptive information than someone might need to find the plant.

Some plants were more useful than others. The plant that Dioscorides called *panax heraklios,* for instance, produced a juice

WHAT IS A PHARMACOPOEIA?

A pharmacopoeia (also spelled pharmacopeia) is a published list of medicines and other health care products. Modern pharmacopoeias list products that are authorized by the government for use. The term *pharmacopoeia* was first used in 1561 and came into general use early in the 17th century.

Lists of recipes for making medicines from plant, animal, and mineral ingredients have existed since ancient times and were used by apothecaries—an *apothecary* is a chemist who makes drugs. In England grocers also made and sold medicines, but in 1606 James I (James VI of Scotland) issued a charter establishing the Society of Apothecaries of London. This society was linked to the Guild of Grocers, but in 1617 a further charter established the Apothecary Guild, recognizing apothecaries as craftspersons independent of and quite separate from grocers. The charter made it illegal for anyone other than a member of the Apothecary Guild to prepare or sell medicines. In 1618 the College of Physicians published *Pharmacopoeiae Londonensis* (London pharmacopoeia), which was a list of the preparations authorized for use by the guild. The recipes contained up to 70 ingredients; a medicine based on a single ingredient was known as a *simple.*

Many countries now issue their own pharmacopoeia. In addition, the World Health Organization, a United Nations agency, publishes *The International Pharmacopoeia* and the Council of Europe publishes *The European Pharmacopoeia* on behalf of 36 European nations plus the European Union.

Mandrake (*Mandragora officinarum*) is seen here on a page from an illustrated Greek edition of *De materia medica* by Pedanius Dioscorides published in Constantinople in the mid-13th century. Dioscorides described sleeping potions made from this plant and used as a surgical anaesthetic. *(The Pierpont Morgan Library/Art Resource)*

that healed ulcers, coughs, ruptures, convulsions, headaches, stomach pains, toothaches, snakebites, and several other ailments. It improved the eyesight when made into an ointment and rubbed on the eyelids, and the juice mixed with honey was a cure for indigestion. The plant was probably galbanum (*Ferula galbaniflora*), a member of the carrot family (Apiaceae) that grows in the Middle East. Dioscorides' *strychnos megas kepaios* was black nightshade (*Solanum nigrum*), a European member of the potato family (Solanaceae). It is poisonous, but Dioscorides recommended treating skin ailments by rubbing the affected part with its leaves and treating earaches and indigestion with a liquor made from boiling the leaves. He described the use of sleeping potions made from opium and mandragora (mandrake) as surgical anaesthetics. The following illustration, from a 10th-century edition of *De materia medica,* shows mandragora. Mandragora, which is also a member of the Solanaceae, contains hallucinogenic compounds that cause delirium, and for many centuries people believed it had magical powers. Its properties were associated with the fact that its taproot is often divided into two parts a little like human legs. When pulled from the ground, legend had it that the plant screamed and the person uprooting it died. The best way to obtain a root safely was to dig a trench around the plant to expose the upper part of the root, then tie a dog to the exposed root. The dog's owner should then run away. The dog would follow and in doing so pull the root from the ground. That would kill the dog but its owner would survive.

Pedanius Dioscorides was born in about 40 C.E. in the city of Anazarbus (now Anavarza) in the Roman province of Cilicia (now Çukurova, Turkey). He may have studied at Tarsus in Asia Minor and Alexandria, Egypt, where he would have had access to the great library. He became a surgeon in the Roman army during the reign of the emperor Nero, who ruled from 54 C.E. to 68 C.E., and his travels with the army through Greece, Italy, Asia Minor (modern Turkey), and southern France gave him ample opportunity to study the medicinal plants and minerals in the territories he visited.

ALEXANDER THE GREAT AND HIS EMPIRE

Aristotle taught Alexander (356–323 B.C.E.) before the prince inherited the throne of Macedon in 336 B.C.E. and began to extend the empire his father had won by conquest. Later, the garden at the Lyceum in Athens came to contain plants contributed by followers of Alexander's armies. Theophrastus based many of his plant descriptions on accounts he obtained from travelers who had visited distant regions of Alexander's empire (see "Theophrastus, the Father of Botany" on pages 4–6). The library at Alexandria, which Dioscorides probably used, opened during the reign of Alexander's general Ptolemy I Soter (ca. 367 B.C.E.–ca. 283 B.C.E.). Ptolemy I was a close personal friend of Alexander and may have been a fellow student taught by Aristotle. He became ruler of Egypt in 323 B.C.E. following Alexander's death. The actual work of planning and supervising the library was delegated to Demetrius of Phaleron (ca. 350–ca. 280 B.C.E.), a Lyceum student of Theophrastus.

Alexander ruled Macedon as regent from 340 B.C.E., while his father, Philip II, led a large Macedonian army that invaded Thrace, the country bordering Macedon in the east. Later, Alexander quarreled with Philip and left Macedon, but in 336 B.C.E. Philip was assassinated and Alexander, aged 20, became Alexander III of Macedon, later called Alexander the Great. As the news of Philip's death spread, people in the countries conquered by the Macedonians saw a chance to seize their freedom, and Alexander had to act. He began by executing all his opponents inside Macedon, then restored Macedonian rule in Greece, Thrace, and Illyria.

Leaving an experienced general Antipater (ca. 398–319 B.C.E.) to maintain order at home, in 334 B.C.E. Alexander led a large force into Asia Minor as the first stage of his imperial expansion. The Macedonians defeated a large Persian army, secured the coast of Asia Minor, and in 333 B.C.E. they fought and defeated another Persian army, this one led by the Persian king Darius III. Alexander's forces then fought their way southward through Syria and Phoenicia. In the spring of 331 B.C.E. the Macedonians entered Egypt. After a few months they returned to Tyre, where they received reinforcements from Europe and headed eastward. They defeated the Persians once more at Gaugemala, to the east of the Tigris River, occupied Babylon, and then captured Susa and the Persian capital, Persepolis, and Alexander was proclaimed king of Asia. The Macedonians then resumed their pursuit of Darius, whom they found dying from wounds inflicted by one of his own noblemen—who was later captured by the Macedonians and punished by Alexander's orders. Pressing eastward into central Asia, they reached the Tian Shan Mountains in western China, where a statuette of a Greek soldier has been found in a burial site.

In 326 B.C.E. Alexander invited the chieftains from what is now northern Pakistan to submit to his rule. Some agreed, but others opposed the Macedonians and were overcome only after intense fighting in which Alexander was wounded. The Macedonian soldiers, who longed to see their homes and families, opposed Alexander's plan to press farther into India and persuaded him to turn back. They reached Babylon, where Alexander died on June 11, 323 B.C.E., in the palace of Nebuchadrezzar II. Historians are uncertain of the cause of his death, but it is most likely to have been from disease. In one more month Alexander would have been 33 years old. As the following map shows, at its peak his empire extended from Macedon in the west to the Indus Valley in the east, and from the southern shores of the Black and Caspian Seas to Egypt. The empire began to disintegrate within a few years of Alexander's death, breaking into many independent kingdoms.

Alexander had been a student of Aristotle's, and although the young man's education was interrupted he was highly intelligent and Aristotle had exerted a strong influence on his development. Aristotle encouraged all his students to observe and investigate the natural world, and Alexander was keenly interested in plants and animals. A large retinue of scholars accompanied him as he led his armies. They measured distances, prepared maps, studied unfamiliar religions, and also the flora and fauna. When they came across unfamiliar plant or animal species, Alexander would send specimens back to Europe, many of them to the Lyceum in Athens, for the attention of Aristotle himself. He gathered what information he could on local herbal treatments and practiced medicine, personally treating some of the injuries and diseases afflicting his soldiers.

In 332 B.C.E. Alexander founded the city of Alexandria on the site of Rhakotis, an Egyptian fishing port. He intended the city to bridge the cultures of Greece and Egypt, but he remained there for only a few months after construction began, and he never returned. After Alexander's death, Ptolemy I managed to have his body brought to the city.

The library at Alexandria was built as a replica of the Lyceum, with a colonnaded walk and gardens. There were lecture theaters, reading rooms, and a communal dining room, and the library was located next to, and in the service of, the Musaeum, which was a temple used as an academic institution—and the origin of the word *museum.* The managers of the library were instructed to collect all the information in the world. Library representatives attended book fairs in Athens and Rhodes, and whenever a ship arrived in the port any books it carried would be seized and copied, the library retaining the originals and returning the copies to the shipmaster. This is the library where Pedanius Dioscorides may have studied. Much of the library was accidentally burned down in 48 B.C.E. during a civil war, and although a branch of the library, known as the daughter library,

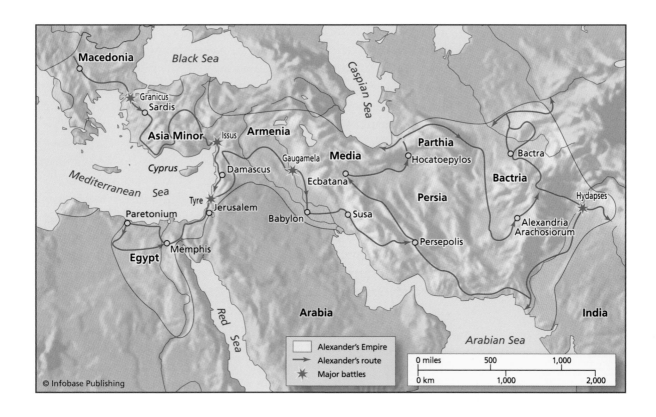

continued to function to the south of the city, it too was destroyed in 391 C.E. when the Roman emperor Theodosius I (347–395) ordered the destruction of all pagan temples.

PLINY, PRESERVING KNOWLEDGE

Knowledge is fragile and easily lost. The books in the library at Alexandria were not bound volumes like those in modern libraries, but texts that were handwritten on scrolls of paper made from the papyrus plant (*Cyperus papyrus*), a type of sedge, and a long book might comprise several scrolls. Later texts, though not those at Alexandria, were written on parchment, which is made from animal skin. Alexandrian merchants had close links with the papyrus producers, which may be why the library preferred papyrus, and the increasing popularity of parchment outside Alexandria may have been due to the heavy demand for papyrus at Alexandria and the consequent rise in its price. In either case, the material is highly flammable.

At its peak, Alexander's empire extended from Macedonia in the west to central Asia and the Indus River in the east, and from the southern shores of the Black and Caspian Seas to Egypt.

The best way to preserve knowledge is to produce many copies and store them in different places. That is simple enough today, but it was more difficult before the invention of printing. An alternative might be to gather as much information as possible and record all of it in a single place, in a book that others could then copy, even though copying meant laboriously writing the entire text by hand. One person who took the latter approach, aiming to gather together as much information on the natural world as he could, was the Roman army officer, administrator, and author Gaius Plinius Secundus (23–79 C.E.), who is better known as Pliny the Elder—"elder" because his nephew and biographer was also called Pliny and is known as Pliny the Younger.

The work in which Pliny recorded everything he could find out about the natural world was entitled *Historia naturalis* (Natural history) and it consisted of 37 volumes—Pliny called them books (*libri*). The first book is a table of contents and a list of all the sources Pliny used. There follow 18 volumes describing nature and a further 18 on practical applications of the knowledge of plants and animals. Books 12 through 17 deal with plants. Books 12 and 13 describe exotic trees—trees that do not grow naturally around the Mediterranean, book 14 deals with vines and wines, book 15 with olives and other fruit trees, book 16 with forest trees, and book 17 with other useful plants such as wheat and barley, including information on storage, milling, making bread, and making porridge. These books also contain detailed descriptions of Roman gardens and gardening methods. Books 20 through 25 and 27 describe drugs obtained from plants.

Pliny was fluent in ancient Greek and was able to translate Greek texts into Latin, but in doing so he claimed Greek knowledge on behalf of Rome, was often disparaging about the Greeks, and in places scholars say his translations were rather too free. Revealing his Roman appropriation of originally Greek material, in the final book Pliny wrote: "Greetings, Nature, mother of all creation, show me your favor in that I alone of Rome's citizens have praised you in all your aspects." That said, the *Natural history* is a valuable summary of the attitudes and state of knowledge in first-century Rome. Pliny is precise and accurate in his descriptions of those plants and their uses of which he had personal experience, drawing heavily on Theophrastus. He had to rely on travelers returning from distant lands for information about plants to which he had no personal access.

Pliny the Elder was born in 23 or 24 C.E. in the city of Novum Comum (modern Como). He had a sister Plinia and a father wealthy enough for him to receive a good education. By the year 30, Pliny was living in Rome as a student. One of his teachers, Publius Pomponius Secundus, had connections at the courts of the emperors Caligula (ruled 37–41) and Claudius (ruled 41–54), through which Pliny was able to embark on a military career when he completed his education. In 45, when he was 21 years old, Pliny went to serve in what is now Germany. He was in Rome briefly in 52, but otherwise remained in Germany. During his time in Germany, he became friendly with the future emperor Vespasian (ruled 69–71) and his son Titus (ruled 79–81). Pliny returned to Rome in 59. Nero (ruled 54–68) was emperor and Pliny maintained a low profile, probably spending most of his time writing. When Vespasian became emperor in 69, Pliny was made a procurator—a government official—with duties that took him through most of the western part of the empire. After some years he returned to the military as prefect of one of Rome's two navies, based at Misenum, on the Bay of Naples.

In August 79 Pliny was at Misenum and his sister Plinia was staying with him together with her son, Pliny the Younger. On August 24, Vesuvius, the large volcano on the opposite side of the bay, became active. Pliny had been out and on his return home he took a bath. It was after his bath that Plinia drew his attention to the cloud above the volcano. Realizing that people were in danger, Pliny ordered the warships to be launched with the intention of using them to evacuate the inhabitants of the towns across the bay. By the time they arrived it was evening. They landed at Stabiae, where Pliny spent the night with his friend Pomponianus. According to his nephew's account, Pliny dined cheerfully, or with the pretense of cheerfulness in order not to alarm his hosts, and then he went to bed.

In the middle of the night, Pliny was roused from his bed. Rocks were falling close to the house, and the building itself was shaking badly. Everyone decided they would be safer in the open, so they left the house, using pillows to protect their heads from falling stones. By this time it should have been daylight, but the volcanic cloud made it darker than the darkest night. They and other parties seeking safety had lamps and torches to help them find their way.

They all decided to go to the shore to see whether they could find safety by sailing out from the coast, but the sea was too rough for them to launch boats. Pliny, who may have been asthmatic, lay down on a linen cloth. Twice he asked for cold water, which he drank. Then they saw flames approaching and there was a smell of sulfur. Two slaves helped Pliny struggle to his feet, but he collapsed at once. He was inhaling ash and found breathing difficult and painful. The situation on the shore must have been chaotic, for Pliny was not found until the morning of August 26. His body was intact, and he looked as if he was asleep.

JOHN RAY AND HIS ENCYCLOPEDIA OF PLANT LIFE

In 1686, 1688, and 1704, the English naturalist John Ray (1627–1705) published the three volumes of *Historia generalis plantarum* (A general account of plants); each volume had approximately 1,000 pages. Ray was 59 and this was the culmination of his life's work, in which he attempted to classify plants. The book described more than 18,600 species, most of them European, and as well as describing them Ray included information on their distribution and ecology, germination, growing habits, diseases, and, where appropriate, their pharmaceutical uses.

Until Ray published his work, the traditional method by which naturalists classified plants involved using a list of plant characteristics. A new specimen would be checked to see whether or not it possessed the first characteristic on the list, then the second, and so on, at each point dividing the possible identification along two routes, so the final identification was achieved by progressively narrowing down the possibilities until only one remained. Ray was the first naturalist to reject this approach and instead to classify plants on the basis of their visible differences and similarities. This led him to place plants into groups sharing many features in common. Although he was not the first naturalist to distinguish between monocotyledons and dicotyledons (see "Theophrastus, the Father of Botany" on pages 4–6), he may have been the first to use this as a major division in his method of classification. Ray recognized the species as the basic unit for classification and was the first naturalist to use the term *species* in its modern sense. His *Historia plantarum* remained a standard botanical textbook in Britain throughout most of the 18th century, and Ray became known as the father of English natural history.

Historia plantarum was not Ray's first botanical work. He had been interested in botany from an early age, and in the 1650s he found himself with the leisure to study plants. He had fallen physically and mentally sick in 1650, and his recovery was slow. It took six years for him to recuperate, and during this time he explored the countryside around Cambridge, where he was living, and grew plants in a garden to which he had access. In 1659, his study completed, he published *Catalogus plantarum circa Cantabrigiam nascentium* (Catalog of plants growing in the vicinity of Cambridge—usually known as the Cambridge catalog). In later years he toured England examining the flora, and in 1670 he published *Catalogus plantarum Angliae et insularum adjacentium* (Catalog of the plants of England and adjacent islands), along the same lines as the Cambridge catalog.

Ray was elected a fellow of the Royal Society in 1667, and in 1673 he submitted to the society a paper entitled "A Discourse on the Seeds of Plants," in which he emphasized the importance of the differences between monocotyledons and dicotyledons. He developed this distinction further in his book *Methodus plantarum nova* (New botanical method), published in 1682. Ray's work was not perfect. He distinguished woody from nonwoody plants without recognizing that these may be closely related, and his listings of plants included a large number of anomalies that he was unable to place. Nevertheless, his was the first serious attempt to produce a system of classification based on natural features reflecting relationships.

John Ray (or Wray as he styled himself until 1670) was born on November 29, 1627, at Black Notley, a village in Essex, on the northern side of the River Thames and to the east of London, where his father was the village blacksmith. His mother was a herbalist and practiced herbal medicine. Ray attended school in Braintree, the nearest town, and when he was 16 he enrolled at Catherine Hall (now St. Catherine's College) at the University of Cambridge. He transferred to Trinity College in 1646 and graduated with a bachelor's degree in 1648 and a master's degree in 1651. Ray became a minor fellow of Trinity College in 1649 and held several college offices. He lectured in Greek in 1651, mathematics in 1653, and humanities in 1655, and one of his pupils was Francis Willughby (1635–72). Ray was ordained a priest in 1660.

In 1660 the English monarchy was restored. Charles II came to the throne, and in 1662 the Act of Uniformity came into force, requiring all clergy to be ordained by bishops and to sign an agreement to use the Book of Common Prayer in Church of England services. This was Charles's attempt to end religious dissension by standardizing the liturgy, the Book of Common Prayer being largely based on the Elizabethan prayer book of 1559. Approximately 2,000 clergymen felt unable to agree, and John Ray was one of them. His refusal to sign meant that Ray had to resign his fellowship at Trinity and leave the university. He had lost his academic career and his livelihood.

Francis Willughby came to his rescue. Willughby, a keen naturalist, was independently wealthy and supported Ray financially. In the spring of 1663 Ray and Willughby, accompanied by two more of Ray's students, set off on a tour of Europe. Ray and Willughby separated at Montpellier, in southern France, Willughby continuing into Spain while Ray returned to England. When Willughby returned, the two began planning a joint work in which they would use the specimens they had collected as the basis for a complete plant and animal classification. The agreement was that Ray would write the volumes on plants, while Willughby dealt with the animals. Ray lived at Willughby's home, Wollaton Hall, Nottingham. In 1672 Willughby died unexpectedly, having completed the work on animals except for the birds and fishes. These were left for Ray to edit. Willughby had bequeathed him an annuity and asked him to tutor his three children, so Ray continued to live at Wollaton Hall. In 1673 Ray married Margaret Oakley, a governess in the Willughby household, and after a time Willughby's widow, Emma, forced the couple to leave. They went first to Sutton Coldfield, Warwickshire, in 1676. In 1677 they moved to Falborne Hall in Essex, and in 1679 they went to live in Black Notley, where Ray remained for the rest of his life. His health deteriorated slowly, but he continued studying and writing until he died on January 17, 1705.

Herbals and
Physic Gardens

It was during the course of the 20th century that the expansion of the pharmaceutical industry supplied physicians with factory-made drugs to treat specified illnesses. Until that time remedies had been based either on therapeutic substances obtained almost entirely from plants—although a few remedies were obtained from minerals or animals—or on interventions such as bleeding, applying leeches, sweating, and purging that were more likely to injure the patient than effect a cure. With the rise of modern medicine, the older interventions fell into disuse, and the use of herbal preparations became increasingly marginalized until they disappeared from official medicinal practice, although medical herbalism continued to thrive informally.

Since plants with therapeutic properties were so important, it is hardly surprising that physicians were required to study botany as part of their training and that close to medical schools there were gardens growing medicinal plants. The art of healing was once called physic, which is why medical doctors are called physicians, and the gardens were known as physic gardens. Such gardens were cultivated in most parts of the world, and, although many became neglected when the medical need for them disappeared, in recent years interest in them has revived and new ones are being planted, although for educational rather than medical uses. In 1965 the Royal College of Physicians in London established a new physic garden. Perhaps the most famous surviving physic garden is at Chelsea, London.

This chapter describes some of the most important physic gardens and also developments in the identification and description of medicinal herbs. Before a gardener can begin cultivating herbs there must be a catalog of those the physic garden should contain. A list of such plants is called a herbal.

The story begins in ancient China with the emperor Shennong. It visits Aztec Mexico and the gardens of Europe, and it ends with an explanation of one of the more curious aspects of the theory of herbalism.

SHENNONG, THE DIVINE FARMER

Traditional Chinese medicine is based largely on herbs, and it has been practiced for a very long time—about 5,000 years. Chinese legend attributes its origin to an emperor who is thought to have lived from about 2737 B.C.E. to about 2697 B.C.E., probably not far from the city of Xian in what is now Shaanxi Province. That emperor was called Shennong. Many long-lived civilizations trace their origin to a time when history and myth are indistinguishable and their kings were divine. Shennong may have existed as a real person, but he was also a god. He was also known as the Emperor of Fire (Yan Di) or the Yan Emperor and the Emperor of the Five Grains. Shennong means "divine farmer" and refers to the belief that he taught his people how to cultivate plant crops and how to use herbs to cure illnesses. He is usually portrayed draped in leaves, with two horns on his head that associate him with the water buffalo, the draft animal used to plow the land—although Shennong is portrayed using a two-pronged spade, a farm implement that preceded the plow. Miniatures of these spades came to be used as charms, and during the Zhou dynasty (11th century B.C.E. to third century C.E.) the Chinese used spade coins, called *bubi*, based on the same shape. The following illustration shows one of these spade coins.

Shennong is credited with having changed the Chinese diet from one based on shellfish, meat, and wild fruit to one based on cereals and vegetables (see "The Story of Rice" on pages 132–134). The change was necessary because the population was increasing and there were insufficient wild animals, but it also meant that people no longer had to kill animals for food. There is archaeological evidence that people were growing both the principal types of cultivated rice (*Oryza indica*

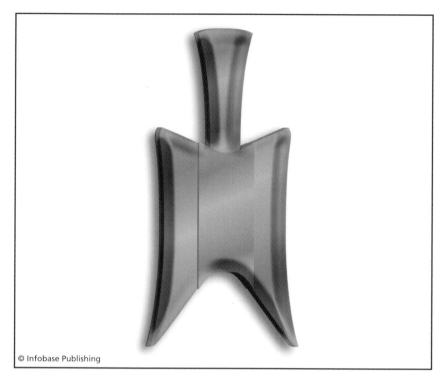

© Infobase Publishing

The earliest Chinese spades, of the type Shennong is often portrayed with, were used to till the land for centuries before the invention of the plow. Charms were shaped like spades, and spade coins, called *bubi,* were used during the Zhou dynasty from the 11th century B.C.E. until the third century C.E. This is a spade coin from that time.

and *O. japonica*) in China by 5000 B.C.E., so by the time of Shennong rice farming was probably expanding into new areas.

In China, food and medicines have always been closely linked. As well as introducing people to plant foods, Shennong investigated the therapeutic properties of herbs by the simple expedient of tasting them. He is reputed to have tasted 70 different herbs on a single day and eventually he tasted all the herbs that grow in China. His stomach was transparent, making it possible to observe what happened to the herbs that he had eaten. He was frequently poisoned, of course, but he discovered a remedy. One day a tea bush caught fire. The rising hot air carried twigs with burning leaves through the air and some of them fell into Shennong's cauldron of boiling water. When the emperor drank the resulting infusion, he found it worked as an antidote to the poisons. In the end, though, he died from eating a type of grass that was alleged to break the intestine. Shennong is also credited with having invented acupuncture.

Shennong described the herbs he tried and listed their properties in a book entitled the *Shennong bencao jing* (Herbal classic of

Shennong). The earliest record of this work contains references to laws that were in operation during the latter years of the Western Han dynasty (206 B.C.E.–220 C.E.) however, so it was clearly compiled many centuries after Shennong's death. The work is unlikely to be authentic, but nevertheless it is the earliest known Chinese pharmacopoeia (see "What Is a Pharmacopoeia?" on page 8), and it describes 365 medicines derived from plants, animals, and minerals.

Together with his close relative the Yellow Emperor Huangdi, Shennong is regarded as one of the ancestors of all the Chinese peoples. He is also regarded as an ancestor of the Vietnamese people.

THE AZTEC HERBAL

Plants grow almost everywhere, many have therapeutic properties, and people of every culture treat their ailments with the herbal remedies they find around them. The peoples of the Americas are no exception, and when Europeans first arrived in Central America they found that the Aztec people were highly skilled practitioners of herbal medicine. The Spanish authorities established schools in their new territories, and in 1552 Martinus de la Cruz and Juannes Badiano (sometimes called Badianus, which is the Latin translation of his name) were Native American students at the Colegio de Santa Cruz in Tlaltilulco. Between them they compiled a list of traditional Aztec herbal remedies. De la Cruz was a Nahua physician and wrote down the remedies he used in the Nahuatl language and drew illustrations. Badiano, an Aztec aristocrat, later translated the Nahua into Latin, with the title *Libellus de medicinalibus indorum herbis* (Little book of Indian medicinal herbs). The result, known as the Badianus Manuscript, is the earliest written American herbal. In the 17th century the manuscript was in the possession of Cardinal Francesco Barberini, so it is also known as the Codex Barberini. The work is now held in the Vatican Library. In 1939 William Gates translated it into English. The original Nahuatl version is lost.

De la Cruz arranged his descriptions according to parts of the body rather than the ingredients used in the remedies. He began with complaints affecting the head, proceeded to those affecting the respiratory and digestive systems, and ended with signs of approaching death.

There is also a work of 12 books known collectively as the Florentine Codex—because it is kept at the Biblioteca Medicea Laurenziana in Florence, Italy. The books are written in Nahuatl, and they were prepared between about 1540 and 1585 under the supervision of Bernardino de Sahagún (1499–1590), a Franciscan missionary who became fluent in the Nahuatl language. Sahagún taught in Tlaltelolco, and he began collecting information about the Aztec way of life and systems of belief. In 1558 the head of his order, Francisco de Toral (1502–71), asked him to write down what he had discovered, believing it would prove useful to those who were instructing and seeking to convert the local people. Accordingly, with the assistance of four of his former students from the Colegio de Santa Cruz who were trilingual in Nahuatl, Spanish, and Latin, Sahagún spent two years interviewing village elders and others and recording the interviews in the Nahuatl language. Between 1575 and 1577 Sahagún translated the Nahuatl text into Spanish and prepared a richly illustrated version with the Spanish and Nahuatl texts side by side. This work is entitled *Historia general de las cosas de Nueva España* (General history of the things of New Spain). Book 11 was devoted to plants, animals, and minerals, and the second chapter describes herbs and their medicinal uses. This is the second longest chapter (more space is given to snakes and other venomous animals), indicating the importance the Aztecs attached to herbal medicine and the extent of their knowledge. Not all of the plants had medicinal uses, but Sahagún's informants told him of 142 that did, and he recorded detailed descriptions of the plants, their habitat, and their therapeutic uses.

The Badianus Manuscript records the methods and prescriptions of one physician, and the work survives only in its Latin translation. The Florentine Codex, based on interviews with a wide number of individuals by an educated Spaniard who was fluent in their language, is considered to be a more authentic account of Aztec medicine and horticulture, although it is not a pharmacopoeia or herbal. Many of the herbal preparations that it describes used fragrant flowers, the Aztecs evidently believing that a perfumed bath was highly beneficial. They had remedies for a range of digestive disorders as well as treatments for gout, fatigue, and arrow wounds.

The Aztecs were keen gardeners. The Nahuatl language has several names for different types of garden and the emperor Montezuma shared the passion. He maintained separate gardens for flowers and medicinal herbs and would not permit fruit or vegetables to be grown in them, saying it was "unkingly" to grow plants for utility or profit in gardens intended for a higher purpose.

ALBERT THE GREAT AND THE STRUCTURE OF PLANTS

Herbalists need to be able to identify plants, and if they are to cultivate them they need to understand the environmental conditions their herbs require. They require the skills of the horticulturist to produce the raw materials from which their pharmaceutical skills allow them to concoct remedies. These are important skills, but they are not the skills of the botanist, who studies the anatomy of plants and the processes of growth and reproduction.

In medieval Europe, botanists relied heavily on ideas and theories that had originally been propounded by Aristotle and Theophrastus (see "Aristotle and His Natural History" on pages 2–4 and "Theophrastus, the Father of Botany" on pages 4–6). So authoritative were these authors that later scholars found it difficult to advance beyond them. Possibly the first scholar to do so was Albertus Magnus (Albert the Great, ca. 1200–80). One of the most learned men of the age, Albert was also called the Universal Doctor. Among his pupils was Thomas Aquinas (1225–74), who was called the Angelic Doctor. Roger Bacon (ca. 1220–ca. 92), a scholar who was an enemy of Albert, was known as the Admirable Doctor.

Albert was familiar with *De plantis* (Of plants), a book that was probably written by the Syrian philosopher and historian Nicolaus of Damascus (born ca. 64 B.C.E.), a close friend of Herod the Great (73–4 B.C.E.) and tutor to the children of Antony and Cleopatra. The writings of Aristotle and Theophrastus first reached Europe in the ninth century in manuscripts based on the work of Isidore of Seville (ca. 560–636), who obtained his information from Pliny (see "Pliny, Preserving Knowledge" on pages 13–16). Nicolaus belonged to a different tradition, of 13th-century writers concerned mainly with Aristotle, who translated the original Greek into their own language, Syriac. Arab scholars then translated the Syriac into Arabic, and European scholars translated the Arabic into Latin or sometimes Greek. *De plantis* was a pseudo-Aristotelian work, which is to say that it had gone through this sequence of translations, although Albert believed it to be the original work of Aristotle.

Having studied *De plantis,* Albert wrote a commentary on it, entitled *De vegetabilibus et plantis* (On vegetables and plants), in which he included many of his own detailed descriptions of plant structures that must have been based on his own observations. He noted, for instance, that a vine sometimes produces a tendril instead of a bunch of grapes, from which he concluded that the tendril is a bunch of grapes that failed to develop. He observed that a thorn is a modified part of a stem and a prickle a superficial structure growing from the outer layer of the stem. Albert believed, with Aristotle, that plants have vegetable souls and questioned whether the souls of two different plants could unite if they lived in close proximity—such as ivy clinging to a tree. He also believed that one species could change into another, as when mistletoe appears on the rotting wood of a dying tree.

Albert, count of Bollstädt, was born in about 1200 in Lauingen an der Donau, Swabia, in what is now Bavaria, southern Germany. He commenced his education either at home or in a local school before enrolling at the University of Padua, Italy, to study liberal arts. Following his graduation in 1223, he joined the Dominican Order in Padua and studied theology at the University of Bologna, before becoming a teacher. He taught theology for several years in Cologne, where the Dominicans had a house, and then in Regensburg, Freiburg, Strasbourg, and Hildesheim. In 1241 he was ordered to go to the Dominican house of Saint-Jacques at the University of Paris, where he taught for four years and in 1245 received a master's degree in theology. Thomas Aquinas was one of his students and in 1248 returned with him to Cologne, where Albert took up the post of regent of the newly established Studium Generale, a Dominican university, and Thomas became the second professor and master of the students. In 1254 Albert was elected head of the Dominican Order in the province of Teutonia (now Germany), a mainly administrative position from which he resigned in 1257 and returned to Cologne. He was appointed bishop of Ratisbon in 1260, but resigned in 1262 and returned to his previous post at the Studium Generale. He died in Cologne on November 15, 1280. Albert was declared a Doctor of the Church on December 16, 1931, making him a saint (Saint Albertus Magnus). In 1941 Pope Pius XII made him the patron saint of natural scientists.

KONRAD VON MEGENBERG AND HIS ILLUSTRATED HERBAL

A modern field guide to plants or a book about the plants found in a particular region would sell rather few copies if it lacked illustrations. Modern photographic and printing technologies make it relatively simple to include colored illustrations, but this is a recent advance. Before the invention of photography and offset printing, an illustration in a book printed from movable type began as carved block made from a very hard wood that would survive the wear it received in the printing press. Each line in the illustration was raised on the surface of the block by cutting away the wood on either side. The woodcut engraver was highly skilled. Consequently, illustrated books cost a great deal to produce and to buy, and illustrations were usually included purely for ornament, to make the book more visually attractive. In that way the printer could produce two versions, one with illustrations and the other without, to sell at different prices.

The first book to be printed in Europe with illustrations as an integral part, augmenting the text rather than merely decorating it, was published by Hans Bämler in Augsburg, Germany, in 1475, and many of the illustrations were of plants. The book was called *Puch der Natur* (The book of nature) and, very unusually for the time, it was in German rather than Latin. Its author was Konrad von Megenberg (1309–74).

Von Megenberg based his work on an earlier book *Opus de naturis rerum* (Work on natural things) of 20 volumes by Thomas of Cantimpré (1201–72), a priest born near Brussels who was a student of Albert the Great in Cologne. Although von Megenberg may have set out to reproduce Thomas's work, in doing so he corrected it, added many of his own observations, and omitted much of what Thomas had written. His aim was to present in one volume a summary of everything that was known about natural history. The book had eight sections called books on:

1. mankind, anatomy, physiology (50 chapters)
2. the sky, the seven planets, astronomy, and meteorology (33 chapters)
3. zoology: quadrupeds (69 chapters), birds (72 chapters), sea monsters (20 chapters), fish (29 chapters), snakes, lizards, and other reptiles (37 chapters), worms (31 chapters)

4. ordinary trees (55 chapters) and aromatic trees (29 chapters)
5. herbs and vegetables (89 chapters)
6. precious and semiprecious stones (86 chapters)
7. 10 kinds of metal (1 chapter)
8. streams and rivers (1 chapter)

There was also a section on races of fantastic humans found in distant lands. Rather more than one-quarter of the book was devoted to plants. *Puch der Natur* was reprinted many times and remained in circulation until the 16th century. More than 100 copies still survive, and all of them are now very valuable.

Von Megenberg was born in 1309, probably at Mainberg, Bavaria, although he calls his birthplace Megenberg. He studied first at the University of Erfurt and later at the University of Paris, where he obtained a master's degree and stayed on for several years teaching philosophy and theology. He was appointed head of St. Stephen's School in Vienna in 1337. In 1342 he moved to Regensburg, Bavaria, to become a parish priest. Later he was promoted to an official ecclesiastical position at the cathedral. He was a highly prolific author who, in addition to his book on nature, wrote on astronomy and physics, economics, history, and moral philosophy, as well as biographies of several saints. He also composed hymns and wrote poetry. He died at Regensburg on April 11, 1374.

CONRAD GESSNER, THE GERMAN PLINY

A polymath is a scholar who is learned in a wide range of subjects, a kind of all-purpose expert. Conrad Gessner (1516–65) of Zürich, Switzerland, was one of the most remarkable polymaths the world has ever known. Some historians have described him as a one-man Royal Society, but during his own lifetime he was best known as a botanist. His ambition was to collect, condense, and then disseminate as much as he could of the information that was accumulating throughout Europe. He became a clearinghouse for information, and it was his passion for this work that earned him his other nicknames—the German Pliny (although he was Swiss) and the father of bibliography. He wrote in Latin, as did most scholars in the 16th century, and after his death this led to some confusion about the spelling of his name. In Latin his name is Gesner, but he always signed his name Conrad Gessner, also spelling his given name with a *C* rather than a *K*. The Latin spelling survives in the plant genus *Gesneria,* belonging to the family Gesneriaceae (African violets and gloxinias), with about 3,500 species.

Gessner and other scholars found themselves compelled to record natural history because by the 16th century they were becoming aware of the limitations of the works of Aristotle and Theophrastus. The Greek authors had been very familiar with the plants that grow around the eastern Mediterranean and the extensive trade with Asia Minor and North Africa had allowed them to learn about the flora of those regions, but they had absolutely no knowledge of the plants and animals living to the north. Consequently, the classical works on natural history available to naturalists living north of the Alps contained no references to many of the species they saw all around them. They had no choice but to describe northern species and to illustrate the works in which their descriptions were published. There was also considerable traffic between Europe and the New World, and American plants were arriving in Europe. These were different from European species and obviously they were not described in the classical texts: *Gesneria* is a North American and Caribbean species. Gessner set out to assemble the information that had been acquired since classical times.

Most of Gessner's botanical works were published posthumously in Nuremberg, between 1751 and 1771, in two volumes entitled *Opera botanica* (Botanical works), which were compiled from his uncompleted manuscripts. In 1541 Gessner published a dictionary of plants *Historia plantarum* (Account of plants) and in 1561 he published *De hortis Germaniae* (On the gardens of Germany). His descriptions of gardens included French, Swiss, and Italian examples and also mentioned individuals who were keen gardeners. Gessner obtained his information by visiting sites that were within reach; for more distant ones he relied on correspondence. He exchanged letters with individuals in France, England, and with two individuals in Poland, but most of his correspondents were in southern Germany, Switzerland, and Italy.

His studies of local plants took Gessner into the Swiss mountains, which he loved. In 1556 he published in Zürich a pamphlet with the title *De raris et admirandis herbis, quae sive quod noctu luceant, sive alias ob causas, Lunariae nominantur* (On rare and admirable plants called Lunariae, either because they glow at night or for some other reason). This contained his personal observations of plants and explanations of the differences between plants with similar names. In it, however, he wrote of his enjoyment of the long walks he took in the mountains. With some friends he had climbed Mount Pilatus, a 7,000-foot (2,120-m) peak beside Lucerne, and he described the pleasure with which they refreshed themselves by drinking from a spring just below the summit and eating bread soaked in the spring water. "I scarcely know if a greater, more Epicurean pleasure (though it is most sober and frugal) can touch the human senses," he wrote. Gessner insisted that in order to gain the most benefit from the experience, a person exploring the mountains must be a true student and admirer of nature.

Gessner also wrote about animals, seeking to separate fact from mythology. His *Historiae animalum* (Accounts of animals) was published in five volumes between 1551 and (posthumously) 1587. The first volume contained 1,100 pages.

Conrad Gessner was born in Zürich on March 26, 1516. His father, Urs, was a furrier. Urs Gessner died in 1531, in the Battle of Kappel, which took place during the religious wars gripping Switzerland. His mother was Agathe Fritz (or Frick). After his father's death money was short, but patrons ensured that Gessner could continue his studies. In 1532 Gessner studied Hebrew at the University of Strasbourg, where he also taught Greek. He then moved to Basel and Paris, but in 1534 the persecution of Protestants meant he had to leave Paris and return to Zürich, where in 1536 he married Barbara Singerin. She had no dowry, and the couple were poor. For a time Gessner had to teach in an elementary school, but friends helped pay for him to study medicine at the University of Basel. In 1537 his patrons helped him obtain the post of professor of Greek at the Academy of Lausanne, where he also had the opportunity to study botany. While at Lausanne, Gessner compiled a Greek-Latin dictionary, which was published in 1539.

The Zürich city physician persuaded Gessner to complete his medical studies, so he visited the University of Montpellier, France, and then moved to Basel, where he qualified as a physician in 1541. After graduating Gessner returned to Zürich as a lecturer in physics at the Collegium Carolinum, which later became the University of Zürich, combining this with a medical practice. In 1554 he was elected city physician, and he remained in Zürich for the rest of his life.

Gessner is known as the father of bibliography because of his *Bibliotheca universalis* (Universal bibliography), published in four volumes between 1545 and 1549. He intended this to be a catalog in Latin, Greek, and Hebrew of every writer who had ever lived, with the titles of all their works and his own comments. He listed approximately 1,800 authors and about 10,000 titles in the first edition. In 1538 he augmented the work with a further listing of 30,000 entries arranged by subject, but this work was never completed. His last work, published in 1555, was *Mithridates: De differentis linguis*, a study of about 130 languages.

In 1564 the Holy Roman Emperor Ferdinand I (1503–64) raised Gessner to the nobility. Gessner died in Zürich on December 13, 1565, while tending the victims of plague.

REMBERT DODOENS AND THE FIRST FLEMISH HERBAL

In 1578 a printer called Gerard Dewes, living at the Sign of the Swan in St. Paul's Churchyard, London, published a new herbal for which he claimed much. He advertised it as (with the original 16th-century spelling):

> A nievve herball, or, Historie of plantes: wherein is contayned the vvhole discourse and perfect description of all sortes of herbes and plantes, their diuers and sundry kindes, their straunge figures, fashions, and shapes: their names, natures, operations, and vertues, and that not onely of those whiche are here growyng in this our countrie of Englande, but of all others also of forrayne realmes, commonly used in physicke/ first set foorth in the Doutche or Almaigne tongue, by that learned D. Rembert Dodoens, physition to the Emperour/ and nowe first translated out of French into English, by Henry Lyte Esquyer.

The book Dewes was offering was the English translation of *Cruydeboeck* (Plant book), a herbal with more than 700 illustrations, written by Rembert Dodoens (1517–85) and first published in 1554. The French translation, *Histoire des plantes*, had appeared in 1557. This was the most extensively translated book of its time and appeared in a total of 13 editions, growing from 877 pages in the first edition to more than 1,500 pages. (It remained the most widely used botanical reference work for more than 200 years.) Copies of Lyte's English translation, as well as the Dutch, French, and Latin editions, are held in several museums and are bought and sold by antiquarian booksellers (they are very expensive!). Henry Lyte (ca. 1529–1607) was a botanist and antiquary.

Dodoens obtained much of his information and some of the woodblocks for his illustrations from the work of the German botanist Leonhard Fuchs (see "Leonhard Fuchs, *Fuchsia,* and the First Botanical Glossary" on pages 38–41), but there was an important difference. Fuchs had arranged his plants alphabetically; Dodoens arranged them in six groups according to their properties, which he considered to be "species, form, name, virtue, and temperament" (by "virtue" he meant "usefulness"). With each translation Dodoens took the opportunity of refining and expanding the work. By the time the Latin translation (*Stirpium historiae pemptades sex*) appeared in 1583, published in Antwerp by Christopher Plantin, the plants were arranged in 26 groups, there were 1,309 woodcut illustrations, and the work came to 900 pages. Effectively, it was a new book.

Rembert Dodoens was born Rembert Van Joenckema on June 29, 1517, in Mechelen (Malines is the French name) in what was then the Spanish Netherlands and is now Belgium. He studied medicine at the University of Louvain, graduating in 1535 when he was 18. He spent some time traveling in France and Germany before returning to his hometown in 1538 where he settled down as the town physician. In 1539 he married Kathelijne De Bruyn(e), and from 1542 until 1546 the couple lived in Basel, Switzerland. Dodoens wrote works on cosmography and physiology before turning to botany with *De frugum historia* (On the natural history of fruit), published in 1552. In 1557 he refused an invitation to become a professor at the University of Leuven. His wife died in 1572, which was also the year the Dutch population rose up against the Spanish occupation. In the course of the revolution, Dodoens's house was looted and the town burned. He lost all of his possessions. The king of Spain, Philip II (1527–98), invited him to become his personal physician but Dodoens declined, becoming instead the physician to the Holy Roman Emperor Maximilian II (1527–76) and his successor, Rudolph II (1552–1612), accompanying them to Vienna and Prague. In 1582

Dodoens returned to the Netherlands, becoming professor of medicine at the University of Leiden, where he remained until his death on March 10, 1585.

JOHN GERARD AND HIS HERBAL

In the 16th and 17th centuries, barbers performed surgery as well as cutting hair and shaving men. The red-and-white barber's pole is believed to represent blood and bandages associated with their merged calling, and in Britain surgeons are still addressed as "Mr." rather than "Dr." This arrangement began in 1163, when a papal decree forbade priests and monks from shedding blood. Up to that time, monks acted as physicians and performed minor surgery, but the decree compelled them to pass their surgical duties over to the barbers. Monks had to be clean-shaven, so every monastery employed one or more barbers.

John Gerard (1545–1612) was a barber-surgeon who lived in Holborn, London, where he had a garden either in the grounds of his house or on ground he leased in nearby Fetter Lane. In 1596 he published a list of more than 1,000 plants that he cultivated in his garden. These included culinary herbs such as Gibraltar mint (*Mentha pulegium*), rosemary (*Rosmarinus officinalis*), saffron (*Crocus sativus*), and thyme (*Thymus vulgaris*), and medicinal herbs such as foxglove (*Digitalis purpurea*), deadly nightshade (*Atropa belladonna*), valerian (*Valeriana officinalis*), and wormwood (*Artemisia absinthum*). Gerard also grew the apothecary's rose, also known as the red rose of Lancaster (*Rosa gallica*). Gerard's list is the earliest catalog of all the plants in a single garden. A copy of it still exists in the British Museum. He was also superintendent of the gardens belonging to William Cecil, Lord Burleigh (1520–98) in the Strand, London, and at Theobalds House, near Cheshunt, Hertfordshire, from 1577 until Burleigh's death in 1598.

When Dodoens's *Stirpium historiae pemptades sex* appeared in 1583, John Norton, a London printer, commissioned Dr. Robert Priest to prepare an English translation. Priest died before completing the task and Norton asked Gerard to take over. He completed the translation, but made a number of alterations to it and added 182 plants and other observations of his own, based on the plants in his own garden, some of which were native to North America and had been given to him by friends. Gerard also added previously unpublished information from Matthias de L'Obel, also called Matthaeus Lobelius (1538–1616), a French botanist living in London who had been physician and botanist to James I and VI. To illustrate the book, Norton had obtained the woodcut blocks that had been used in the *Eicones plantarum seu stirpium* by the German botanist Jacob Theodorus Tabernaemontanus, published in Frankfurt in 1590. A further 16 woodcuts were added to Gerard's book. These included the first illustration of the potato (*Solanum tuberosum*), captioned "Potatoes of Virginia" to distinguish them from the sweet potato (*Ipomoea batatas*). He grew potatoes in his garden. At that time they were considered a great delicacy and only the rich could afford them. Gerard also included some mythical plants, including the barnacle-goose tree—"The breede of Barnakles"—which he claimed to have seen and informed his readers that it could be found on an island in Lancashire.

There was a folk belief that goose barnacles (*Lepas anatifera*) began life as growths that appeared on logs floating in the sea, so they were plants, and that as they grew they changed into barnacle geese (*Branta leucopsis*). Gerard described what he believed to be the tree that produces barnacles. It was a persistent myth. Gerald of Wales (ca. 1146–ca. 1223) repeated it in his book *Topographia Hiberniae* (Description of Ireland). Frederick II (1194–1250), the Holy Roman Emperor, was an authority on falconry and birds in general. He sent an expedition to northern Europe to check the story, which he thought dubious. Frederick's envoys returned with barnacles clinging to rotten wood, but these were quite unlike any bird. Albert the Great (see "Albert the Great and the Structure of Plants" on pages 24–26) performed an even more thorough test. With some friends, Albert bred a barnacle goose with a farmyard goose and found the resulting eggs hatched into perfectly ordinary goslings, establishing beyond doubt that barnacle geese reproduce in exactly the same way as other geese. Nevertheless, the belief was a long time dying, and Gerard helped prolong its decline.

Gerard's work appeared in 1597 with the title *The herball, or, Generall historie of plantes gathered by John Gerarde of London, master in chirurgerie.* In the preface Gerard described the work as "the first fruits of these mine own labours" and included the following rather curious reference to Dr. Priest: "Doctor Priest, one of our London

Colleagues hath (as I heard) translated the last edition of Dodonaeus, which meant to publish the same; but being prevented by death, his translation likewise perished." Gerard's book described more than 2,800 plants and contained approximately 2,700 illustrations. It had three volumes. Volume 1 covered grasses, rushes, reeds, cereal grains, irises, and bulbs (all of which are monocotyledons). Volume 2 described plants valued for food, medicine, or their sweet smell. Volume 3 dealt with roses, trees, bushes, shrubs, plants grown for their fruit, plants producing gum and resin, heaths and heathers, mosses, and fungi. The work also had an index. Thomas Johnson, a London apothecary, corrected and expanded Gerard's work, publishing the result as a second edition in 1633. This was reprinted in 1636.

John Gerard was born in Nantwich, Cheshire, in 1545. He was educated locally and then became a ship's surgeon until 1562, when he moved to London and apprenticed himself to a barber-surgeon. After completing his seven-year apprenticeship, he was permitted to set up his own practice. This was successful, and Gerard became well respected. He became a member of the Court of Assistants in the Barber-Surgeons Guild in 1595 and subsequently held various offices. In 1608 he became the Guild's Master. In 1596, the year he published his garden catalog, the Barber-Surgeons Guild commissioned Gerard to create a "fruite-grounde" for their London premises, which he did. Gerard died in London in February 1612 (in the modern calendar, 1611 in the old calendar when the year began on March 25).

NICHOLAS CULPEPER AND HIS HERBAL BEST SELLER

In 1653 Nicholas Culpeper (1616–54) published *The Complete Herbal*. It was the last of his works to be published in his lifetime, and he wrote the following explanation of how he felt obliged to offer an alternative to what he regarded as the unnatural and harmful treatments of his medical rivals.

> This not being pleasing and less profitable to me, I consulted with my two brothers, DR. REASON and DR. EXPERIENCE, and took a voyage to my Mother NATURE, by whose advice, together with the help of DR. DILIGENCE, I at last obtained my desire; and, being warned by MR. HONESTY, a stranger in our days, to publish it to the world, I have done it.

The *Complete Herbal* had first appeared in 1652, entitled *The English Physitian, or an Astrologo-physical discourse of the vulgar herbs of this nation. Being a complete method of physick, whereby a man may preserve his body in health; or cure himself, being sick.* Culpeper's aim was to supply the information that would allow ordinary people to identify medicinal plants they could find growing in fields and hedgerows near their homes and use them to make ointments, poultices, infusions, or decoctions that would heal them. Sometimes he described precisely where a particular herb could be found. The book sold for three pence, which made it accessible to the poor, and Culpeper wrote in simple, straightforward English that ordinary people could understand. *The Complete Herbal* was an immediate success. For more than 250 years it was the principal guide to traditional cures that ordinary people used, and it is still in print. Nicholas Culpeper is regarded as one of the founding fathers of modern herbalism.

Both the book and its author were highly controversial, however. Culpeper firmly believed in astrology. Medicine had been strongly influenced by astrology since classical times, so Culpeper was not introducing anything new, but times were changing. Astronomy had become a science, and astronomers had shown astrology to be mere groundless superstition. That was not the only reason the medical establishment attacked Culpeper, though. He had struck the first blow, and the physicians were defending themselves.

Medical practitioners had to belong to the Royal College of Physicians, and they were allowed to prescribe only those remedies that were published in the *Pharmacopoeiae Londonensis* (London pharmacopoeia), published by the Royal College and dispensed by members of the Society of Apothecaries. The *Pharmacopoeiae Londonensis* was written in Latin, and physicians wrote their prescriptions in Latin. Most patients had no idea what medicines they were being prescribed. It was a secretive closed shop, and to compound matters doctors and apothecaries were charging exorbitant prices.

Culpeper did the unforgivable: in 1649 he translated the *Pharmacopoeia Londonensis* into English and published it as *The London Dispensatory*. He gave his reasons in the following words.

> I am writing for the Press a translation of the physicians' medicine book from Latin into English so that all my fellow countrymen and

apothecaries can understand what the Doctors write on their bills. Not long ago parsons, like the predecessors of my grand-father William Attersole, used to preach and pray in Latin, whether his parishioners understood anything of this language or not. This practice, though sacred in the eyes of our ancestors, appears ridiculous to us. Now everyone enjoys the gospel in plain English. I am convinced the same must happen with medicine and prescriptions.

His translation revealed that the ingredients going into the officially authorized medicines included many different kinds of feces, the skull of a man who had met a violent death, the brain of a sparrow, the horn of a unicorn, the fat of a lion, and many that were even more disgusting and even more absurd. The Royal College denounced him and said he was endangering lives by encouraging untrained readers to self-medicate, but they could not prevent publication or punish the author. In 1641 Parliament had abolished the Star Chamber. Until that time anyone who offended the sovereign—and the Royal College was under the sovereign's patronage and therefore protected—committed an offense that would be tried in the Star Chamber, which had the power to impose severe penalties and seize and destroy offensive publications. As an alternative to the official list of medicines, on the title page of *The Complete Herbal* Culpeper offered: "Such things only as grow in England, they being most fit for English bodies." What was even worse from the physicians' point of view was that Culpeper's remedies were free.

Nicholas Culpeper was born on October 18, 1616, at Ockley, Surrey, where his parents, the Reverend Nicholas Culpeper and his wife Mary Attersole, had moved the previous year. The Reverend Culpeper died when his son was 13 days old, and the infant was raised in Isfield, Sussex, by his grandfather, the Reverend William Attersole, an intellectual but also a stern and devout Puritan. Attersole planned that Nicholas should study at the University of Cambridge and enter the church. From an early age, however, Nicholas read avidly, from books in his grandfather's library, about astrology, medicine, and medicinal plants.

In 1632, when he was 16, Culpeper entered Cambridge, but the courses on offer did not appeal to him, and he spent his time enjoying himself. He also took up the new habit of smoking. He and his childhood sweetheart, Judith Rivers, had planned to marry. Judith was a wealthy heiress whose family would never have permitted such a marriage, so the couple decided to elope, arranging to meet near Lewes, Sussex, where they could find a ship to take them to the Netherlands until the fuss died down. On the way to the rendezvous, Judith's coach was struck by lightning and she was killed. Devastated, Nicholas abandoned his studies. Culpeper's mother died a year later, and his grandfather disinherited him, outraged because he refused to return to Cambridge. Instead, the Reverend Attersole paid £50 for him to be apprenticed to Daniel White, an apothecary in London, and Culpeper began his seven-year apprenticeship in November 1634. Before completing the apprenticeship, however, White went bankrupt and disappeared to Ireland, taking what was left of Culpeper's fees with him. Another apothecary, Francis Drake of Threadneedle Street, agreed to take the apprentice on, Culpeper teaching him Latin as payment for his board and lodging. About 18 months later, Drake died. Culpeper and a fellow apprentice, Samuel Leadbeaters, took over the business, working under supervision until they had finished their training.

In 1639 Culpeper married Alice Field, the 15-year-old daughter of a wealthy merchant and a former patient. Marriage meant he had to abandon his apprenticeship—apprentices were not permitted to marry. He had completed five years and felt he knew enough to go into practice. He and Alice built a house and shop in Red Lion Street, Spitalfields, which was a poor district of London, and he opened a business as an astrologer, physician, and herbalist. The Society of Apothecaries denounced him for not being fully qualified, and physicians disapproved of the fact that Culpeper treated the poor, often for free.

Civil war broke out in 1642, and Culpeper, an ardent Puritan and republican, enlisted, serving as a field surgeon. On September 20, 1643, a musket ball hit him in the chest while he was tending a casualty. He was carried back to London and never fully recovered. He and Alice had seven children, but only their daughter Mary survived. In 1651 Culpeper published *A Directory for Midwives*.

Nicholas Culpeper died on January 10, 1654. The combination of tuberculosis, his old wound, and overwork was given as the cause of his death, but his secretary suggested that his heavy smoking might also have contributed. At the time of his death he had 79 books or translations awaiting publication.

LEONHARD FUCHS, *FUCHSIA*, AND THE FIRST BOTANICAL GLOSSARY

It is the custom for botanists to honor distinguished members of their profession by naming plants for them. The fuchsia is a popular ornamental shrub, grown for its flowers and as a hedging plant. The French botanist Charles Plumier (1646–1704) was the first European to discover it. In 1696–97 Plumier found it growing wild on the Caribbean island of Hispaniola (occupied by the Dominican Republic and Haiti). Fuchsia can also be found in tropical and subtropical Central and South America and in New Zealand and Tahiti. The cultivated plant is one of about 100 species in the genus *Fuchsia* belonging to the evening primrose family (Onagraceae). The cultivated plant was introduced to Europe and North America, and it has escaped from cultivation and become naturalized in many places. The following illustration shows fuchsia growing in a hedge in Ireland. Fuchsia is also the name given to the typical color of the flower—although there are cultivated varieties with other colors. Plumier named the fuchsia in honor of the German physician and botanist Leonhard Fuchs (1501–66), who was the first botanist to compile a botanical glossary; he called it *Fuchsia triphylla, flore coccineo.*

Fuchs was trained as a physician and in the 16th century, when most medicines were herbal, botany was an essential part of the medical curriculum. But Fuchs went further than most medical students. He accepted that while a knowledge of plants was necessary for a physician, the study of plants should be enjoyable in itself. He derived great pleasure from walking through the woods and meadows and observing the great variety of plant life, and he believed that anyone studying plants would delight in learning to recognize and understand them. In 1542 Fuchs published a book about plants. Its title was *De historia stirpium commentarii insignes* (Notable commentaries on the history of plants). The book appeared first in Basel, Switzerland, and in 1543 it was translated into German, Dutch, and English (its English title was *New Herbal*).

This book was different from all the herbals that had gone before. It was customary to include illustrations of plants in such a work, but Fuchs believed that the illustrations should be very precise and accurate, so they could be an aid to identification. In the preface to his work he wrote the following.

The genus *Fuchsia* is made up of more than 100 species of plants originally from Central and South America. They are now widely cultivated for their attractive flowers and as hedges, and they have become naturalized in Europe. These fuchsias are growing in a hedgerow near the town of Dingle, in County Kerry, Ireland. *(Dennis Flaherty/Photo Researchers, Inc.)*

As for the pictures themselves, every single one of them portrays the lines and appearance of the living plant. We were especially careful that they should be absolutely correct, and we have devoted the greatest diligence that every plant should be depicted with its own roots, stalks, leaves, flowers, seeds, and fruits.

The llustration (page 40) of common plantain (*Plantago major*) based on the woodcut in the *New Herbal* shows how detailed Fuchs insisted the drawings should be, and in order to provide as much botanical information as possible, some of the illustrations showed flowers, fruits, and seeds all on the same plant at the same time. That is a botanical impossibility, but it is a valuable technique for maximizing the utility of the drawing. Fuchs had invented botanical illustration, and he insisted that artists work under his direct supervision to ensure botanical accuracy. Three artists contributed to the work, and the book includes a picture of all three of them working from a plant in a vase. Albrecht Meyer drew the illustrations, Heinrich Füllmaurer transferred Meyer's drawings onto woodblocks, and Veit Rudolf Speckle carved the blocks. There is no use of shading in the

© Infobase Publishing

This reproduction of an illustration from *De historia stirpium* shows how detailed the botanical drawings were in Fuchs's book. In order to show all the parts of the plant, leaves are pushed aside so the plant appears spread out. This plant is *Plantago major*.

drawings, as the illustration of the plantain shows. That is because Fuchs intended the illustrations to be colored and the book appeared in both colored and black-and-white editions.

Fuchs was inspired by an earlier work, the three volumes of *Herbarium vivae eicones* (Pictures of living herbs) by another German botanist Otto Brunfels (1488–1534) that was published between 1503 and 1506. Brunfels illustrated his herbal with pictures drawn from living plants, some pictures showing withered or insect-damaged flowers and leaves. But Fuchs was more ambitious. The *New Herbal,* which took him more than 30 years to produce, described approximately 400 wild plants and more than 100 cultivated plants, and it had 512 pictures, many drawn from specimens he had grown in his own garden. The plants were listed alphabetically by their Greek names and they included some New World species that Fuchs was describing for the first time, including corn (*Zea mays*) and chili pepper (*Capsicum* species—Fuchs called it *siliquastrum*—"big pod"). Not all Fuchs's descriptions were accurate. He relied on classical authors, especially Dioscorides (see "Pedanius Dioscorides and His Catalog of Medicinal Plants" on pages 7–9), but Fuchs had never visited the Mediterranean and sometimes confused characteristics of Mediterranean plants with those of northern European plants, which were not the same. He also listed some species more than once under different names, because of mistaken identification. Nevertheless, his *New Herbal* was a magnificent achievement.

The *New Herbal* was written for the benefit of physicians. Fuchs complained that it was difficult to find even one physician in 100 who had accurate knowledge of even a few plants. It was to help them that he included a glossary—perhaps the first—of botanical terms.

Leonhard Fuchs was born on January 17, 1501, in Wemding, Bavaria. He was educated at a school in Heilbronn and from the age of 12 at the Marienschule in Erfurt, Thuringia. He qualified as a physician in 1524 at the University of Ingolstadt, Bavaria, and practiced medicine in Munich from 1524 to 1526, when he became professor of medicine at Ingolstadt. In 1528 he became personal physician to Georg, margrave of Brandenburg, remaining in this post until 1531. In 1533 he moved to the University of Tübingen, where he was chancellor several times and for many years professor of medicine. Fuchs remained at Tübingen until his death on May 10, 1566.

THE BAUHIN FAMILY

Herbals grew steadily longer in the course of the 16th century, and as they grew so did the number of omissions and inaccuracies they contained and repeated. There was only one way the situation could be remedied: Someone had to go through as many of the publications as possible looking for duplications, synonyms, and obvious errors, and comparing descriptions and illustrations with plant specimens. The Swiss botanist Caspar (or Gaspard) Bauhin (1560–1624) spent some 30 years patiently doing exactly that. In doing so, he replaced the discursive, almost rhapsodic descriptions his predecessors had employed with extremely brief, factual accounts of flowers, stems, leaves, and roots that would aid in classification.

In 1596 Bauhin published the first results of his effort, as *Phytopinax* (Plant images). This was only partially completed at the time of its publication, but it proved popular, and Bauhin followed it with *Pinax theatri botanici* (Theater of botanical images), published in 1623 and consisting of 12 books with a total of 72 sections in which he described more than 6,000 plants. This work was a concordance—an alphabetical listing of plant names used by other authors—and it contained no illustrations. On the title page Bauhin described it as an "index to the works of Theophrastus, Dioscoridea, and the botanists who have written in the last century." He classified plants as trees, shrubs, and herbs and placed the spice plants in a category he called Aromata. The names Bauhin used were descriptive, but he defined plant species and grouped species in genera, and in many cases his descriptive name for a genus or species was reduced to a single word. He ended each entry with the name, often abbreviated, of the individual who first used it. For example, he named one of the species of bluegrass, also called fescue, as Festuca prior, Dod. (Dodoens). This being a concordance, he also listed all of the synonyms. Superficially his scheme resembles the binomial system of nomenclature that Linnaeus used (see "Carolus Linnaeus and the Binomial System" on pages 83–86), and Linnaeus was certainly influenced by it, but there is an important difference. Bauhin used names purely as descriptions, whereas Linnaeus used them as unique identifiers. A description must refer to some feature of the species by which it can be identified. An identifier need not describe the species or even have a meaning, but it must be a name that applies only to that species and not to any other. Thus, although many authors claim that Bauhin invented the binomial system that Linnaeus later adopted, this is not so.

Bauhin also compiled a catalog of the plants growing in the area around Basel, and he planned another major work, *Theatrum botanicum* (Theater of plants). He intended this to be a huge 12-volume work, but he had completed only three of the volumes by the time of his death, and only one volume was published, in 1658.

Jean (or Johann) Bauhin (1541–1613), Caspar's elder brother, was also a botanist. His major work, *Historia plantarum universalis* (Universal history of plants), was a summary of everything known about plants at that time, with descriptions of more than 5,000 species. He died before completing it, but it was finished by his son-in-law Jean-Henri Cherler and published in 1650–51 at Yverdon, Switzerland.

The Bauhin brothers were the sons of Jean Bauhin (1511–82), a French physician from Amiens who was forced to leave France because of his Protestant faith and who settled in Basel, Switzerland. Caspar Bauhin was born in Basel on January 17, 1560. He studied medicine at Padua, Montpellier, and at several schools in Germany. He qualified as a physician in Basel in 1580, and for a time he taught botany and astronomy privately. He was appointed professor of Greek at the University of Basel in 1582, and in 1588 he became professor of anatomy and botany. Later he was appointed official physician to the city of Basel and professor of medicine and later rector and dean of the medical faculty at the university. He died in Basel on December 5, 1624.

Jean Bauhin was born in Basel on December 12, 1541. He studied botany first at the University of Tübingen under Leonhard Fuchs (see "Leonhard Fuchs, *Fuchsia*, and the First Botanical Glossary" on pages 38–41) and later at the University of Zürich under Conrad Gessner (see "Conrad Gessner, the German Pliny" on pages 27–30). He accompanied Gessner on botanical excursions in Switzerland, and Gessner thought very highly of him. Bauhin then began to practice medicine in Basel, and in 1566 he was appointed professor of rhetoric at the University of Basel. In 1570 he became personal physician to Duke Frederick I of Württemberg, who lived in Montbéliard, in eastern France. He was also director of the Montbéliard botanic gardens, which were among the oldest in Europe. While there he developed the technique for growing potatoes. Bauhin remained at Montbéliard until his death on October 26, 1612.

MONASTIC GARDENS

In the Middle Ages European monasteries and convents were centers of learning, and the monks and nuns also had a duty to care for the lay community outside the monastery walls. They provided shelter and hospitality for pilgrims and other travelers, schools for local children, and they cared for the sick. Most monasteries and convents observed the Rule of St. Benedict. This prescribed a life with four principal components: work, study, leisure, and prayer. The physical work demanded of monks and nuns contributed to the upkeep of the community, and tending the gardens and farmland formed an important part of it, because a religious house had to be self-sufficient, in medicines as well as basic foodstuffs, wines, and ales.

A monastery or convent would have orchards to supply tree fruits and gardens growing nonwoody plants. Religious houses as far north as northern England also maintained vineyards. During what climate historians call the medieval warm period, global average temperatures were markedly warmer than those of the early 21st century, and England was a major producer of high-quality wines.

Monastic gardens were typically designed as a rectangle broken into smaller rectilinear areas by straight paths, and there would be two distinct gardens: a physic garden and a kitchen garden. As the name suggests, the physic garden or herb garden grew medicinal plants and culinary herbs. A member of the community who specialized in herbal remedies would tend the medicinal plants, gathering ingredients at the optimum time and using them to make therapeutic preparations. The kitchen gardens supplied food for the community.

The usual arrangement was that a path ran along the center of a kitchen garden and the ground on either side of the path was divided into at least three beds, making a minimum of six beds in all. Fences or *hurdles*—temporary fences, often made from woven willow, hazel, or ash, which could be moved as required—defined the boundaries of the beds. A physic garden would have many more beds, one for each of the herbs it contained. At the center of the garden, where the paths crossed, there would be a well to supply the community with water. There was also a cloister garden, comprising an area of lawn, sometimes bordered by flowerbeds. The cloister was where monks and nuns would spend time in contemplation.

All monastic gardens also possessed fishponds, moats, and streams. These contained carp and eels to supply food on days when it was forbidden for the community to eat meat.

Gardens of this type were to be found throughout Christendom, and people were very familiar with them. When it became fashionable for aristocrats to have gardens outside their houses, they based their designs on monastic gardens, but gradually they made innovations that modified the central concept. In *De vegetabilibus* (About vegetables), published in 1259, Albert the Great (see "Albert the Great and the Structure of Plants" on pages 24–26) described a pleasure garden that he must have visited. Essentially this remained close to the monastic model, but it is possible to detect the beginning of a change. The pleasure garden had beds growing herbs and vegetables and beside them a meadow with a fountain and grassy banks on which people could sit, and there were trees to provide shade. The emphasis was shifting and the medieval pleasure garden later developed into a walled garden that existed to provide enjoyment as well as to supply food and herbs.

THE APOTHECARIES' GARDEN AT CHELSEA

In every major city there are annual events that mark the passing of the seasons. London is no exception, and one of its annual festivals attracts gardeners and garden enthusiasts from all over Britain. The Chelsea Flower Show is a celebration of garden design and plant breeding. The show takes place in a garden that has existed since 1673, when it was established by the Worshipful Society of Apothecaries on the site of an existing garden belonging to the politician Sir John Danvers (1588–1655). It was known originally as the Apothecaries' Garden.

Apothecaries prepared and dispensed medicines, most of which were herbal, and medicinal plants were grown for the purpose in physic gardens. Since part of their job was to pick, or supervise the picking of, the plants they required, apothecaries had to be able to identify plants reliably and recognize when the plants were in peak condition. They learned their profession through a seven-year apprenticeship, and the Apothecaries' Garden at Chelsea is where they learned botany. During the 18th century, it was the most richly stocked botanical garden in the world. It is now called the Chelsea Physic Garden and covers 3.5 acres (1.4 ha), although it was once larger.

Chelsea is a district beside the Thames and to the west of central London. The garden is located close to the river, where the *microclimate*—the atmospheric conditions found in a small local area—is sufficiently mild for plants from warmer climates to thrive. The garden contains the largest fruiting olive tree (*Olea europaea*) growing outdoors in Britain (protected by a wall), the world's most northerly outdoor grapefruit, and the world's oldest garden rockery growing alpine plants. In 1700 the garden began exchanging seeds with Leiden Botanical Garden in the Netherlands, inaugurating an ongoing international scheme to exchange seeds with other botanic gardens. In 1712 Sir Hans Sloane (1660–1753; see "Sir Hans Sloane, Milk Chocolate, and the British Museum" on pages 144–146) purchased what was then the Manor of Chelsea and leased the site of the garden to the Society of Apothecaries for an annual rent of £5 on condition that each year the garden supply the Royal Society with 50 good herbarium samples of up to 2,000 plants. A *herbarium* is a collection of preserved plant specimens or the building in which such a collection is housed. The garden is now run as a nonprofit organization. The following illustration shows plants growing in one of the garden's greenhouses.

The Apothecaries' Garden was established as an educational facility, and botanists still train there. The garden is also open to the public, who are welcome to walk its paths and examine its plants—enjoying an informal kind of education. What visitors see in the pharmaceutical garden is a display of the plants a medieval herbalist might have picked in a monastic garden arranged in the way that would have been familiar to a monk or nun physician.

Its educational purpose influenced the garden's design. Almost since their inception (see "Pisa, Padua, and Florence, the First Botanical Gardens" on pages 62–63), botanical gardens have set out their plants in *order beds.* These are beds in which all the plants belong in a particular category. For example, Caspar Bauhin (see "The Bauhin Family" on pages 42–43) categorized plants as trees, shrubs, and herbs, so a botanical garden might have beds or areas devoted to each of these. Alternatively, each bed might hold plants that flower in a particular month or plants producing particular types of fruit (nuts, berries, drupes [plums are drupes], etc.), or types of flowers. There are many ways to plan an order bed, but nowadays they are usually based on an accepted system for botani-

Chelsea Physic Garden, in London, has existed since 1673. Originally it was a place where students could learn to identify medicinal plants. This is greenhouse 56. *(Peter Jousiffe/Science Photo Library)*

cal classification, so all the plants in a bed might belong to the same family or (if it is large) genus. Many of the order beds in the Chelsea garden follow a late 19th-century taxonomic arrangement. In years to come this may be replaced by a more modern system of classification.

Chelsea's medicinal plants are arranged differently, in order beds related to their uses rather than *taxonomy*—biological classification. There are beds for British plants used in a variety of medical specialties such as dermatology, ophthalmology, parasitology, and cardiology. There is also a garden of world medicine, with beds arranged geographically with plants used in Native American, Maori, Aboriginal, and other traditional medical systems.

THE DOCTRINE OF SIGNATURES

People are very good at recognizing patterns, even where no pattern exists, and pattern recognition played an important part in early herbal medicine. When people rummaged through the meadows and along the hedgerows for plants that might heal their ailments, they saw similarities between plant forms and the part of the body that was sick. In time the existence of these similarities acquired a theological justification.

In the 16th century, the Swiss physician, alchemist, and astrologer Paracelsus (1493–1541) challenged many of the accepted medical views of his day. He called himself Paracelsus—beyond Celsus—to show his superiority to the Roman encyclopedist Aulus Cornelius Celsus (ca. 25 B.C.E.–ca. 50 C.E.), whose book *De medicina* (About medicine) summarized Roman medical knowledge. Paracelsus's real name was Phillip von Hohenheim and he also called himself Philippus Theophrastus Aureolus Bombastus von Hohenheim. He spent much of his life traveling as a vagabond with a group of supporters, but eventually he settled down as professor of medicine at the University of Basel, where he dramatically burned several of the existing textbooks. Paracelsus attacked the greed of apothecaries and quarrelled with the medical establishment. He believed that diseases were caused by external factors rather than imbalances within the body, and he aimed to discover a cure for every complaint. He was a very successful doctor, but highly unpopular with the authorities.

Paracelsus also maintained that God had left clues hinting at the true purpose of all the objects he had created and that the skill of the physician depended on understanding the divine clues relating to the treatment of disease. For example, Paracelsus taught that pricking the patient with a thistle was an effective treatment for internal inflammation. He pointed out that the seeds of skullcap (*Scutellaria*

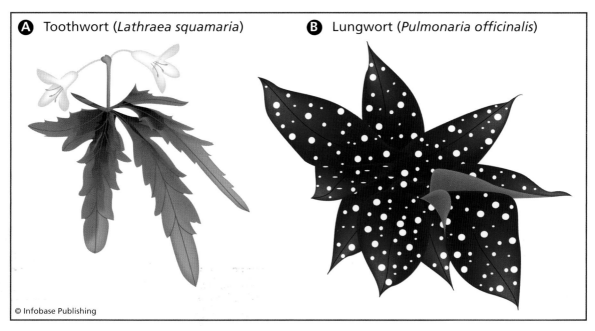

Ⓐ Toothwort (*Lathraea squamaria*)   Ⓑ Lungwort (*Pulmonaria officinalis*)

© Infobase Publishing

Left: Toothwort (*Lathraea squamaria*) leaves have edges suggestive of a row of teeth. Right: The blotches on lungwort leaves (*Pulmonaria officinalis*) are reminiscent of the surface of lung tissue.

species) resembled skulls—hence the name—and that the plant cured headache. Willow (*Salix* species) grows in damp places and cures rheumatism, which is associated with dampness. It was not a new idea, and the uses of plants were often reflected in their names. For example, snakeroot (*Ageratina* species) was supposed to be an antidote to snake venom, toothwort (*Lathraea* species) to relieve toothache, and liverwort (*Hepatica* species) to cure liver disorders. The illustration above shows the features of toothwort and lungwort on which the similarities were based.

This is the teaching that came to be known as the doctrine of signatures, the signatures being the marks God had used to label plants. The German theologian and mystic Jakob Böhme (1575–1624) developed and popularized the doctrine in his book *Signatura rerum* (The signature of all things), published in 1622, and in the years that followed the supposed links grew. People believed that long-lived plants promoted longevity, plants with yellow sap cured jaundice, eyebright (*Euphrasia* species) was good for eye ailments, lousewort (*Pedicularis* species) repelled lice, and there were many more. The English herbalist William Coles (1626–62) mentioned in his book *The Art of Simpling*, published in 1656, that walnuts "bear the whole signature of the head," and wrote of viper's bugloss (*Echium vulgare*) that "its stalks all to be speckled like a snake or viper, and is a most singular remedy against poison and the sting of scorpions." Nicholas Culpeper (see "Nicholas Culpeper and His Herbal Best Seller" on pages 34–38) also promoted the doctrine.

These beliefs were not confined to Europe. People in most cultures have associated the colors, shapes, and markings of plants with parts of the human body. Obviously there can be no true links of this kind and supposed similarities between plants and organs of the body are purely coincidental. The doctrine of signatures is pure superstition with no rational basis.

Botanical Gardens and Herbaria

Physic gardens grew medicinal plants and as well as supplying the plants they were used to teach botany to apothecaries and physicians. In time botanists came to appreciate the value of growing collections of nonmedicinal plants as well. Plants were arriving in Europe from the New World, Africa, and Asia that European botanists had never seen before. By the middle of the 16th century, physic gardens were expanding to become botanical gardens and, in parallel with that development, lists of useful herbs were becoming lists of all the plants found in a particular area. This chapter traces those changes.

The chapter begins by explaining how the study of herbs expanded to become the study of all plants. It tells of the establishment of botanical gardens and the assembling of the world's great botanical collections. It was not possible to cultivate every plant species, but it was possible to preserve plants in other ways, and the chapter outlines the rise of the herbarium. Private ornamental gardens were also becoming fashionable, and the chapter describes that development and what it aimed to achieve.

IDENTIFYING PLANTS: THE HERBAL BECOMES THE FLORA

In 1622 Caspar Bauhin (see "The Bauhin Family" on pages 42–43) published *Catalogus plantarum circa Basileam sponte nascentium*

(Catalog of plants occurring naturally around Basel). This slim book was a list of the plants growing around Basel. It contained no illustrations, but it listed all the synonyms for the plants it described and details of where each plant might be found, and alternate pages were left blank for students to make their own notes. The catalog was intended for the use of medical students, but it was not a herbal in the traditional sense. It was possibly the first *flora*—a description of all the plants, useful or not, growing in a specified area.

Botany was still being taught as a branch of medicine, and physicians would continue to study plants until the expansion of the pharmaceutical industry displaced herbalism from the center of medical practice, but botanists were beginning to devote increasing amounts of time to the study of plants in general. Bauhin was a professor of anatomy as well as botany, and his catalog, written late in his career, was based on many years of field excursions. In it he drew attention to medicinal plants, but he did not confine himself to them. He described plants of all kinds. Bauhin's most important work was his *Pinax theatri botanici,* published in 1623, but in 1620 he had published a shorter introductory work, *Prodromus theatri botanici* (Introduction to botanical exposition), in which he included descriptions of 600 plants that no one before him had described, possibly because they had no known medical uses.

Other floras followed in subsequent years. These were local, describing the plants growing close to where the author lived. In 1659 John Ray published his Cambridge Catalog, describing the plants growing around Cambridge (see "John Ray and His Encyclopedia of Plant Life" on pages 16–18). *Plantae Coldenhamiae* by the Scots-born physician, farmer, botanist, and four-time governor of New York Cadwallader Colden (1688–1776), published in 1743 with a revised edition in 1751, was possibly the first American local flora. Colden's flora described the plants on his Coldenham Estate near Newburgh, New York. Linnaeus published Colden's work in Uppsala and spoke highly of it (see "Carolus Linnaeus and the Binomial System" on pages 83–86).

European explorers were also compiling floras of other continents. One of the most notable of these was the Polish botanist and Jesuit missionary Michal Piotr Boym (ca. 1612–59). In 1643 Boym set out on a journey to eastern Asia and described the plants, animals, peoples, and customs of all the lands he visited along the way. He wrote a description of the flora and fauna of Mozambique that reached Rome but was never published. His Chinese flora, on the other hand, was published in Vienna in 1656. Boym illustrated his *Flora Sinensis* in color with annotations in Latin and Chinese (with guides to pronunciation) and, despite the title, he included animals—for some reason including a hippopotamus!—as well as plants. It was a small work, describing only 21 plants and eight animals and with a total of 23 pictures, but it is the earliest book about the plants of subtropical China written by a European.

FORMAL GARDENS, RESTORING ORDER TO A CHAOTIC WORLD

The rising interest in plants was accompanied by a wish to cultivate them, and private gardens became increasingly popular from early in the 16th century. Borrowing the basic design of the medieval monastic garden (see "Monastic Gardens" on pages 44–45), these gardens were usually walled and were tranquil places where people could relax, enjoying the shapes, colors, and perfumes of the flowers.

At first, it was only the wealthy and powerful who could afford these gardens. They were laid out in the grounds of palaces and stately homes, and they came to reflect the artificiality of courtly manners. The following illustration shows a Tudor garden at Hampton Court Palace in London. The palace is situated on land that in the 13th century belonged to the Knights Hospitalers of St. John of Jerusalem, and in 1514 Thomas Cardinal Wolsey (ca. 1471–1530) obtained a 99-year lease on the site from the Knights Hospitalers. Wolsey transformed what had been a stately home into a magnificent palace, where foreign dignitaries would be entertained and impressed by the opulence, and important negotiations would be conducted. As well as being a cardinal, Wolsey was lord chancellor of England, and in the 1520s he failed to persuade the pope to grant Henry VIII (1491–1547) the divorce from Katherine of Aragon (1485–1536) that the king needed in order to marry Anne Boleyn (1501 or 1507–1536). Wolsey fell from favor, and in 1528 he lost Hampton Court, which then became the property of Henry. The Tudor garden was replanted in the 18th century, but it is in the style of palace gardens during the Tudor period, which lasted from 1485 to 1603. It is, literally, a garden in which Henry VIII once strolled.

The Tudor garden at Hampton Court Palace, London. The beds were replanted in the 18th century. *(Jim Steinberg/ Photo Researchers, Inc.)*

Beautiful though they are, these grand gardens were not an attempt to recreate the biblical Garden of Eden; that would have been blasphemous. Nor were formal gardens of this kind meant to be natural in the sense of imitating natural landscapes. The countryside was where food, fibers, and industrial raw materials were produced, and it was aesthetically pleasing only insofar as it was productive. Hunting was popular with the aristocracy, but it took place in carefully managed hunting forests and the animals the hunters killed went to the kitchens; hunting was a type of food production as well as a sport. Beyond the farms and the hunting forests, the wilderness was an unruly and dangerous place. A garden demonstrated not a love of nature, but the total mastery of nature. It was where nature, represented by plants, was carefully positioned and strictly controlled. *Parterres*—extensive level areas with flower beds—were fashionable, in which low box hedges, clipped to a severely rectangular cross section, defined geometrically shaped beds growing arrangements of colorful flowers separated by straight paths.

Those ordinary citizens who were prosperous enough to own a town house with ground around it also designed and cultivated gardens. The designs were based on the parterres and lawns of the grand gardens, and in a very real sense they were the precursors of the modern suburban garden. Formal gardens also contained stone sculptures—another fashion that endures, though now with figures made from concrete or plastic.

Topiary—the art of clipping trees and shrubs to make sculptured shapes—was also popular. It had long been practiced in China and Japan, but European topiary began in Roman times and fell out of favor until it was revived in the 16th century. By the end of the 17th century, all grand gardens and most smaller town gardens had examples of lovingly clipped and trained plant sculptures. Its popularity began to fade when the English poet and satirist Alexander Pope (1688–1744) decided he could stand it no longer. In his 1713 *Essay on Verdant Sculpture,* Pope wrote the following:

Adam and Eve in yew;
Adam a little shattered by the fall of the tree of knowledge
    in the great storm;
Eve and the serpent very flourishing;
The tower of Babel, not yet finished;
St. George in box; his arm scarce long enough, but will be in
    condition to stick the dragon by next April; and
a quickset hog, shot up into a porcupine, by its being forgot a
    week in rainy weather.

In the same essay Pope also wrote:

A citizen is no sooner proprietor of a couple of yews, but he entertains thoughts of erecting them into giants, like those of Guildhall. I know of an eminent cook who beautified his country-seat with a coronation dinner in greens, where you see the champion flourishing on horseback at the end of the table, and the queen in perpetual youth at the other.

Nothing could withstand such withering ridicule, and topiary fell out of fashion. But the whole concept of the garden was changing. Pope was an enthusiast for the new style of landscape gardening.

A wide terrace separated a formal garden from the house and the garden was at a lower level, so someone standing at a window or on the terrace had a view of a long sweep of parterre and lawn, extending into the distance. To enter the garden it was necessary to descend a flight of steps. The new style was to bring the garden right up to the house, remove the geometrical flower beds, and instead give the entire area over to grass, with meandering paths, clumps of trees, small lakes, and views that revealed themselves only when the visitor turned a corner. A new profession of landscape gardener emerged to cater to the new fashion, one of the most famous practitioners being Capability Brown (see the following sidebar).

LANCELOT "CAPABILITY" BROWN

During the 18th century there was a reaction against the excessively formal gardens to be found in the grounds of most European palaces and great houses. A more romantic approach became fashionable, and landowners sought to surround their homes with parkland that appeared natural. Rectangular beds and straight paths gave way to curved lines. A person wandering through the park encountered different views that focused on features such as classical ruins, imitation temples, or small lakes. Small groups of trees were scattered informally, often on top of artificial hills, and lawns extended almost to the door of the house. The overall effect was meant to imitate the countryside, but it was a strictly controlled and visually appealing countryside. Creating and maintaining parkland of this kind was known as landscape gardening.

There were many landscape gardeners earning a good living by designing parks for wealthy individuals. Lancelot Brown was one of the most successful and famous. He designed more than 170 gardens, many of which still exist, and some are open to the public. The designs of other famous gardens, including Kensington Gardens, Kew Gardens, and Hyde Park in London, were influenced by Brown's design style, although he did not design them personally. His style was characterized by gently rolling grassland, clumps of trees, and curving lakes that he made by damming small rivers.

When a landowner invited him to devise a scheme for landscaping a garden, Brown would walk around the estate, looking and thinking, and finally he would announce that he could "see the great capabilities" of the area. This habit earned him the nickname of Capability Brown.

Lancelot Brown was born in 1716 (the date is unknown) in the village of Kirkharle, Northumberland, in the northeast of England, and educated at a school in the village of Combo, two miles (3 km) away. He left school when he was 16

LUCA GHINI AND HOW TO PRESS FLOWERS

Gardeners were often avid plant collectors. Highly competitive, they sought to impress their rivals by their displays of the latest fashions in plantings and of plants that had only recently arrived in their corner of the world. The landscaped gardens with which 18th-century landowners delighted their houseguests were often stocked with exotic trees, and the owners of formal gardens liked to show off the latest blooms, occasionally paying extortionate sums for their plants (see the sidebar "Tulipomania" on pages 67).

Gardens were meant to impress as well as to delight the eye. They were not, however, intended to be museums that preserved plant

and became an apprentice gardener at Kirkharle Tower, the home of Sir William Loraine. In 1739, his apprenticeship completed, Brown moved to Wotton Underwood, the Buckinghamshire estate of Sir Richard Grenville, and the following year to Stowe, the estate belonging to Grenville's son-in-law, the politician Lord Cobham. At Stowe, Brown worked under the distinguished landscape gardener William Kent (ca. 1685–1748). Brown rose through the gardening hierarchy to become head gardener. As head gardener, one of his duties was to accompany visitors as they walked through the gardens. He would explain the principles of landscape gardening and the possibilities it afforded, at the same time revealing his own extensive knowledge. He was soon in demand. With Lord Cobham's approval, Brown designed the redevelopment of the grounds on the adjacent Wakefield Estate of the duke of Grafton. While still working for Lord Cobham, Brown designed several other gardens, including that at Warwick Castle.

Lord Cobham died in 1749, and Brown moved to Hammersmith, now a district in west London, but then a village, where he established a private practice. In 1764 he was appointed master gardener at Hampton Court Palace and moved into an official residence beside the palace, but he was not happy at having to maintain the formal gardens. He bought an estate of his own in 1767, at Fenstanton, a village in Cambridgeshire. Brown died in London on February 6, 1783. Although many of his gardens survive, the landscape gardeners who followed him found his style somewhat bland, and his designs went out of fashion.

By the end of the 18th century, however, even the "natural" landscaped gardens were seen as contrived and bland. Gradually the gardens of large houses came to consist partly of formal flower gardens, though these were less rigidly geometric than those of earlier times, and partly of parkland and managed woodland.

diversity by growing plants for educational or research use or for what would now be called conservation. Those functions were performed by botanical gardens, but even the largest botanical garden could not aspire to grow specimens of every plant species—far less varieties of those species. They lacked the space.

The solution to this difficulty was to hold plant collections in the form of dried and pressed plants—herbaria. Many plants can also be stored as seeds. A *seed bank* is a store where plant seeds are maintained at a constant temperature of 32°F (0°C) and a *relative humidity*—the amount of water vapor present in the air as a percentage of the amount required to saturate the air at that temperature—of about 4 percent (which is extremely dry). Many seeds remain viable for up to 20 years under these conditions, and the seed bank can be made permanent by sowing seeds toward the end of their storage period, growing the plants, and harvesting fresh seeds to restock the bank. Not all plants can be conserved in this way. *Recalcitrant seeds* do not survive being dried and germinate rapidly; most tropical rain forest plants produce recalcitrant seeds and can be preserved only as growing specimens. The possibility of using seed banks to store plants was not discovered until early in the 20th century, however; the Russian geneticist Nikolai Vavilov established one of the first (see "Nikolai Vavilov and the Origin of Cultivated Plants" on pages 158–162).

Herbaria suffer no such limitation. They preserve entire plants, and they have been in use since early in the 17th century. Herbarium specimens are prepared by a method that has remained largely unchanged since that time. A plant to be stored in this way was picked, then laid out on a sheet of absorbent paper in such a way as to display its important features, and a label placed beside it giving the name of the plant, usually in Latin and the botanist's native language. A second sheet of paper was placed on top to cover it, and then other plants were added, each one arranged between two sheets of paper, to form a small stack of specimens. The stack was then placed in a press, where it was held under pressure until it had thoroughly dried. The following illustration shows how simple a plant press can be. Finally, the specimens were carefully removed from the press, separated from their backing sheets, and glued or sewn onto a sheet of paper or board. The sheets might then be bound into a volume containing plants of a particular type.

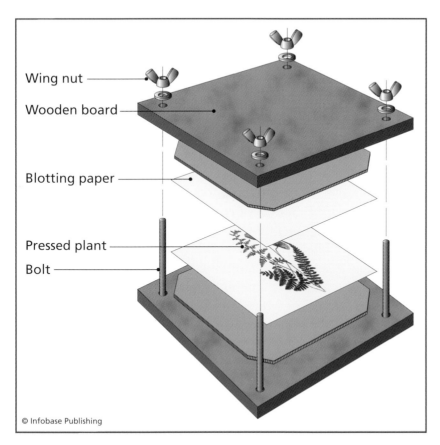

Wing nut

Wooden board

Blotting paper

Pressed plant

Bolt

© Infobase Publishing

Herbarium specimens are dried and pressed in a single operation by laying each plant between two sheets of absorbent paper such as blotting paper. Several plants are processed together as a stack. They can be pressed between two boards linked by four bolts and wing nuts. The press should be placed somewhere warm to accelerate drying.

The Flemish anatomist and botanist Adriaan van den Spiegel (1578–1625) was probably the first scholar to describe how to make a herbarium, complete with a recipe for a suitable glue. This was contained in his book *Isagoge in rem herbarium libri duo* (Introduction to matters relating to the herbarium in two books), the two volumes of which were published in 1606 and 1608. The person who popularized herbaria, however, was the Italian physician and botanist Luca Ghini, the director of the botanical gardens at Pisa and Florence (see "Pisa, Padua, and Florence, the First Botanical Gardens" on pages 62–63). Ghini published no books, but he was the inspired teacher of several eminent Italian botanists, including Andrea Cesalpino (1519–1603), Pietro Andrea Mattioli (1501–77), and Ulisse Aldrovandi (1522–1605). Ghini taught his students to preserve plants in this way, and they passed on his method to their colleagues and students. In

1532, one of his former students, the botanist and artist Gherardo Cibo (1512–1600), assembled a herbarium that still exists.

Luca Ghini was born in 1490 in Imola in the Italian province of Bologna, in the northeast of the country. His father was a lawyer. Ghini studied medicine at the University of Bologna, and in 1527 he became a lecturer there in medicinal botany. In 1539 the university established a chair in that subject, and Ghini was the first professor, but he remained subordinate to the professor of medicine to such a degree that he was unable to obtain permission to create a university physic garden. In 1543 Cosimo I de'Medici (1519–74), the grand duke of Tuscany, invited Ghini to create a botanical garden at the University of Pisa. Ghini moved to Pisa in 1544, and in the same year he made his first herbarium—although that is not what he called it. In the 16th century it was called *hortus siccus* (dry garden) or *hortus hiemalis* (winter garden); the term *herbarium* was introduced in the late 18th century. Ghini returned to Bologna in 1554 and died there on May 4, 1556.

THE RISE OF THE HERBARIUM

There was nothing novel in the practice of pressing and drying plants, especially flowers. The oldest example of a plant preserved in this way consists of the twigs and leaves of an olive tree (*Olea europaea*) made from a bundle of olive twigs bound together by cords made from palm leaves that was discovered in 1885 in an Egyptian tomb. The plant material was placed in the tomb in about 305 B.C.E. and the herbarium specimen is held by the Royal Botanic Gardens, Kew, London. Travelers had been pressing and drying plants for centuries as souvenirs of their travels, much as modern tourists take and keep photographs. Pilgrims also picked flowers along their routes and, once dried between the pages of a Bible or book of devotions, these could serve as objects for meditation as well as reminders of their journeys. What was novel was the idea of preparing plants in this way for use as botanical reminders. Herbaria were, and are still, used to confirm the identities of plants collected in the field. A large modern herbarium will contain several million specimens. The world's largest, at the Muséum national d'Histoire naturelle, in Paris, holds 9.5 million plants, and the second largest, the herbarium at the New York Botanical Garden, has 7.2 million.

botanical garden included teaching and Cluyt and his son Outgaert gave outdoor lessons in summer and indoor tuition in winter. Even after he had recovered Clusius took little part in the running of the garden, spending his time in his own garden where he grew bulbs, and especially tulips.

Carolus Clusius is the Latin name of Charles de l'Écluse or l'Escluse, who was one of the most influential botanists of the 16th century. He was born in Arras, now in northeastern France but then in Flanders, on February 19, 1526, the son of Michel de l'Escluse, who was a councillor at the Artois provincial court. He studied law at the University of Louvain and briefly at Marburg, but in 1551 he

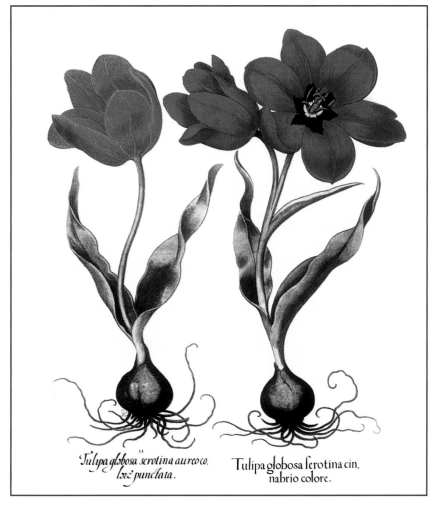

Tulipa globosa serotina aureo co. lxie punctata.

Tulipa globosa serotina cin. nabrio colore.

Although they were colorful and attractive, the first tulips to be seen in Europe were much less showy than tulips were to become after many generations of selective breeding. This illustration appeared in *Hortus Eystettensis* (Eystett's garden) by Basilius Besler (1561–1629), published in 1613. *(Georgette Douwma/Photo Researchers, Inc.)*

was studying medicine and botany at the University of Montpellier, where he qualified as a physician. However, he was more interested in botany and never practiced medicine. During his student years, Clusius became proficient in eight languages, and his first book was a French translation of Rembert Dodoens's *Cruydeboeck* (see "Rembert Dodoens and the First Flemish Herbal" on pages 30–32).

Clusius held a number of posts in different parts of Europe before in 1573 being appointed court physician and superintendent of the imperial garden of the Holy Roman Emperor Maximilian II (1527–76), in Vienna. This position allowed him to travel widely, studying and collecting plants. After moving to Leiden, Clusius remained there until his death on April 4, 1609.

Tulips (*Tulipa* species) grow naturally around the Mediterranean and from Turkey eastward to China. The ancestor of the modern cultivated tulip originated in Turkey and was introduced to Europe in the middle of the 16th century. Tulips grow from bulbs, and Ogier Ghislain de Busbecq (1520 or 1521–92) sent Clusius some bulbs from Turkey, and Clusius took them with him to Leiden. Busbecq was a herbalist and also the ambassador to the Ottoman Empire of the Holy Roman Emperor Ferdinand I (1503–64). The following illustration shows the tulips that were being cultivated at about that time.

Clusius planted his bulbs, and the plants flourished. He observed them closely and discovered how to produce varieties with multicolored and feathered petals. These showy blooms made tulips so popular that the price of bulbs rose rapidly. This soon led to an extraordinary episode of price inflation known as tulipomania (see the following sidebar).

As well as introducing the tulip to northern Europe, Clusius popularized several other garden plants, especially alpine species, but also including the potato. He produced one of the first studies of the flora of Spain (1576) and of Austrian and Hungarian mountain plants (1583). These were published together as *Rariorum plantarum historia* (History of rare plants) in 1601. Clusius's book *Exoticorum libri decem, quibus animalium, plantarum, aromatum, aliorumque peregrinorum fructuum historiae describuntur* (Ten books of exotic life-forms, the history and uses of animals, plants, aromatics, and other natural products from distant lands), published in 1605, was a major botanical and zoological survey.

TULIPOMANIA

The tulip originated in Turkey. It was introduced into Europe in the 16th century and became highly popular in the United Provinces (now the Netherlands). Growers discovered they could produce a wide range of varieties with many different colors, some a single but vivid red or yellow, some white, and some with more than one color. These more exotic multicolored varieties arose because of a *mosaic virus*—a virus that caused the infected plant to produce petals or leaves with specks or patches of color.

Tulips grow from bulbs, and as the fashion grew for displaying the flowers, demand for the bulbs increased. The virus that produces the most desired multicolored flowers propagates through the bulbs, and it took several years to produce bulbs that would reliably deliver the required blooms. As demand increased, the price rose and, starting in 1634, the Netherlands was consumed by a bulb-buying frenzy that was later called the tulipomania. Within the space of two or three months, the price of a single bulb of the variety Admirael de Man rose from 15 guilders to 175 guilders and a bulb of the variety Generalissimo went from 90 guilders to 900 guilders. The most valuable variety, Semper Augustus, sold for 5,500 guilders per bulb in 1633 and in January 1637 it was allegedly worth 10,000 guilders. In 1635, some people were spending as much as 100,000 guilders to buy 40 bulbs. People told their friends about the prices they had paid for bulbs or that they had heard of others paying and as the stories circulated no doubt they improved and the numbers grew larger. Nevertheless, there was certainly a rapid inflation in tulip prices. To place these in context, in 1642 Rembrandt was paid 1,600 guilders for his most celebrated painting, *The Night Watch,* the annual earnings of a carpenter were about 250 guilders, and in 1636 French brandy sold for 60 guilders a gallon (13 guilders per liter).

The end of the frenzy began on February 3, 1637, in Haarlem. Florists seeking to auction tulip bulbs found there were no bidders. Prices collapsed until by May bulbs were selling for between 1 percent and 5 percent of their peak prices.

No one knows what caused the sudden rise and fall in the price of tulips. Many economists maintain that it was not a true bubble, because tulips were genuinely in demand, and the demand exceeded the supply. The national economy was recovering rapidly from a depression that followed war with Spain, and Haarlem suffered a severe epidemic of bubonic plague in the 1630s. These may have been contributing factors, but the true cause of the tulipomania remains a mystery.

SIR HENRY CAPEL, PRINCESS AUGUSTA, AND THE ROYAL BOTANIC GARDENS AT KEW

Britain's principal botanical garden occupies more than 300 acres (121 ha) beside the River Thames at Kew in southwest London. It became a botanical garden in 1752, so what is now the Royal Botanic Gardens, Kew, is of much more recent origin than most of Europe's botanical gardens and it is not the oldest botanical garden in Britain; Oxford Botanic Garden was founded in 1621.

Kew Gardens consists of many specialized gardens and two out-stations that together probably contain the world's largest and most diverse collection of living plants. It is also a partner in the Millennium Seed Bank Project that aims to store seeds from 10 percent of all the wild plant species in the world, and its herbarium holds about 7 million specimens, including about 350,000 holotypes. In 2003 the United Nations Educational, Scientific and Cultural Organization (UNESCO) designated Kew Gardens a World Heritage Site. The following illustration shows that much of the area is laid out informally. The large glasshouse near the top of the picture is the Temperate House, the largest surviving Victorian glass structure.

In 1762 a pagoda in the Chinese style was constructed in Kew Gardens. This photograph of the gardens was taken from the top of it in July 2003. The gardens cover about 300 acres (121 ha) and contain the world's largest collection of living and preserved plants. It is a World Heritage Site. *(Getty Images)*

Although Kew Gardens is today a center for botany and plant conservation, in the second half of the 19th century the emphasis was a little different. The gardens had been in decline during the early 19th century and by 1831 they were continuing to receive plants but were no longer sending out plant hunters to collect them. That situation began to change in the 1840s. Kew Gardens began to expand its scientific research, sending seeds, plants, and advice to the colonies, and when a rail line linked Kew to central London the gardens began to attract visitors. New scientific knowledge was accumulating rapidly, and there was great interest in putting it to practical use. Britain then controlled a very large empire, and the application of botanical knowledge amounted to identifying the plants that could be grown most profitably in a particular colony and ensuring that they were planted there. Kew was used to store, grow, study, and develop plants with commercial potential that could then provide the basis for enterprises in the colonies.

That is not how the gardens began, however. They were originally a pleasure garden. In the 17th century, Kew Park belonged to Sir Henry Capel, a public servant who held important offices, including Lord Justice of Ireland (1693) and Lord Deputy of Ireland (1695). In 1692 he was promoted to become lord capel of Tewkesbury. Kew Park was his house and around it Capel created a garden of exotic plants. Frederick, prince of Wales (1707–51), later leased the house, renamed it Kew Palace, and spent much of his time there. The palace was demolished in 1802, but an adjacent building bought by George III (1738–1820) is now called Kew Palace.

After Frederick's death his widow, Princess Augusta of Saxe-Gotha (1719–72), the dowager princess of Wales, inherited the palace and its garden—the city of Augusta and Augusta County, Georgia, are named in her honor. Augusta greatly expanded Capel's 10-acre (4-ha) garden. Her written instructions to the head gardener were to: "compleat all that part of the Garden at Kew that is not yet finished in the manner proposed by the Plan and to keep all that is now finished." John Stuart, the earl of Bute (1713–92), a former prime minister and friend of Frederick, advised Augusta to have a garden containing all the plants on Earth. The attempt to achieve this marks the point at which Kew Gardens began to become a true botanical garden, and 1752 is given as the official foundation year of the Royal Botanic Gardens, Kew.

Bute also recommended the Scottish architect Sir William Chambers (1723–96) to design the buildings. Chambers disliked the style of landscape gardening developed by Lancelot Brown (see the sidebar "Lancelot 'Capability' Brown" on pages 56–57), who had designed nearby Richmond Gardens for George III, and he sought to reflect the excitement educated people were experiencing as explorers described to them the architecture and cultures of Asia. Chinese styles, known as chinoiserie, were especially fashionable in the middle of the 18th century, and it was in 1762, during Augusta's ownership of the gardens, that the imitation octagonal, 10-floor Chinese pagoda was completed, originally with a Chinese dragon at each corner of the roof. Kew Gardens retain the royal in its official name, but it is now controlled by a board of trustees on behalf of the British government's Department for Environment, Food, and Rural Affairs.

SIR JOSEPH BANKS, UNOFFICIAL DIRECTOR OF KEW

Princess Augusta died in 1772. Her son, George III, inherited both Kew Gardens and Richmond Gardens, and in 1802 he joined them together. George III was keenly interested in agriculture and agricultural research—an enthusiasm that earned him the nickname "Farmer George"—and he turned part of the combined gardens over to raising sheep. During his reign Britain was beginning to industrialize at home and was expanding overseas, while at the same time the country was fighting the Napoleonic Wars. The following map shows the extent of the British Empire in 1822. Although the area of the territories Britain controlled was nowhere near what it would become by the end of the century, nevertheless the colonies were widely scattered geographically and were located in many different climatic zones. Times were changing, and the role of the botanical garden changed with them. Instead of exhibiting exotic plants, Kew Gardens started to conduct serious research, and the naturalist who did most to promote that change was Sir Joseph Banks (1743–1820).

Banks held no official position at Kew, but he was a close friend of the king and shared the king's vision of using the gardens to investigate and develop commercial uses for plants that could benefit the colonies. Banks was highly qualified for this task. He had sailed on several expeditions to collect plants, including the 1768–71 round-the-world voyage of HM Bark *Endeavour* led by James Cook

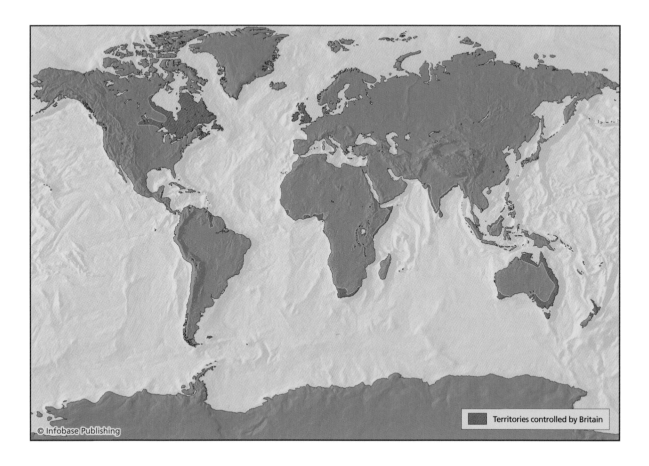

Territories controlled by Britain

© Infobase Publishing

(1728–79), and in 1778 he was elected president of the Royal Society, a post he held until 1819.

During his time at Kew Gardens, Banks organized expeditions to collect plants in South Africa, Ethiopia, India, China, and Australia, and he arranged for plants to be shipped from the gardens to various colonies. It was Banks who arranged the transfer of breadfruit from Tahiti to the West Indies on HMS *Bounty* (see "Captain Bligh, HMS *Bounty,* and the Breadfruit Trees" on pages 138–141). Banks was so successful that by about 1800 almost every ship sailing to Britain from any British colony carried plants destined for Kew, and when new plants reached Kew the gardeners worked to have them on public display ahead of any other European garden.

Banks was born in London on February 13, 1743. His father, William Banks (1719–61), was a wealthy landowner with an estate in Lincolnshire and a Member of Parliament. Bank entered Harrow School

The colored areas indicate territories controlled by Britain in 1822. At that time the British Empire was still expanding, but the territories were widely scattered and located in many different climatic zones.

in 1752 at the age of nine, and in 1756 he enrolled at Eton College but was ill after a smallpox inoculation during the 1760 summer vacation and did not return to Eton at its end. Instead he entered Christ Church College, University of Oxford, in December 1760 but left in 1764 without taking a degree. On his father's death in 1761 Banks inherited the Lincolnshire estate and moved to London, dividing his time between London, Oxford, and Lincolnshire. He continued his botanical studies at Chelsea Physic Garden and made friends with many of the leading scientists of the day, including Linnaeus (see "Carolus Linnaeus and the Binomial System" on pages 83–86). He was elected a fellow of the Royal Society in 1766 and in the same year made his first botanical excursion overseas, to Newfoundland and Labrador. His descriptions of the flora and fauna of that region established his scientific reputation and secured his appointment to Cook's *Endeavour* expedition. In March 1779 Banks married Dorothea Hugesson, and the couple settled in a house in Soho Square, London, where Banks lived for the rest of his life. He was knighted in 1781. In his later years Banks was a trustee of the British Museum. He died in London on June 19, 1820.

SIR WILLIAM HOOKER, THE FIRST OFFICIAL DIRECTOR

Both Joseph Banks and George III died in 1820, and in the years following their deaths interest in the gardens waned and they came close to being abandoned. The new king, George IV (1762–1830), had little interest in them. The decline was arrested with the arrival of the first director, Sir William Hooker (1785–1865), in 1841. The following illustration shows Hooker at about this time.

Banks had recruited a number of young botanists as plant collectors. One of them was his protégé William Hooker, who had been elected a fellow of the Linnean Society in 1806 at the age of 21. Banks sponsored Hooker for an all-expenses-paid expedition to Iceland in 1809. Unfortunately, the ship caught fire on the return journey and all of Hooker's notes and specimens were destroyed. Banks offered him the unpublished notes from his own 1722 expedition, but Hooker's memory was excellent and he was able to write a full report, *Journal of a Tour in Iceland in the Summer of 1809*. This established his scientific reputation, and in 1812 Hooker was elected a fellow of the Royal Society. Later, Banks invited Hooker to collect plants in Java, but Hooker declined. This did not affect the friendship between the two men, and in 1820 Banks helped Hooker to obtain the post of Regius Professor of Botany at the University of Glasgow, a position he held until he moved to Kew in 1841. He was a popular lecturer and as well as his teaching and research commitments he established the Royal Botanic Institution of Glasgow and Glasgow Botanic Garden.

William Jackson Hooker was born in Norwich on July 6, 1785, into a wealthy family. He was educated at a school in Norwich, and after leaving school he devoted himself to travel and to studying natural history, at first with a special interest in ornithology and entomology, but turning later to botany. Hooker spent most of 1814 on botanical excursions through France, Switzerland, and northern Italy. He married in 1815 and settled at Halesworth, a small town in Suffolk, where he developed a large herbarium that acquired an international reputation.

Sir William Hooker (1785–1865), the English botanist who in 1841 became the first director of the Royal Botanic Gardens, Kew. *(George Bernard/Science Photo Library)*

An authority on *cryptogams*—plants that reproduce by spores rather than seeds, including algae, lichens, mosses, and ferns—in 1818 Hooker published *Muscologia Britannica*, which was an account of all the mosses of Great Britain and Ireland. Between 1818 and 1820, he followed this with *Musci exotici* (Exotic mosses), published in two volumes, describing non-British mosses and other cryptogams. Among his many other publications, Hooker also compiled *Flora Scotica*, published in 1821, *British Flora* (1830), and the five-volume *Species Filicum* (Species of ferns) between 1846 and 1864. At Kew, Hooker's principal contribution was the renovation of the buildings, redesign of the landscape, and the establishment of the herbarium. Hooker died from a throat infection on August 12, 1865.

William Hooker's son, Sir Joseph Dalton Hooker (1817–1911), attended his father's lectures at Glasgow from the age of seven. He was educated at Glasgow High School and graduated in medicine from Glasgow University in 1839. His medical qualification allowed Hooker to sail to the south magnetic pole as assistant ship's surgeon on HMS *Erebus* as part of the expedition led by Sir James Clark Ross (1800–62). He succeeded his father as director at Kew. Joseph Hooker was director from 1865 to 1885. He continued to develop the landscape of the gardens and supervised the building of the Temperate House. His most important contribution, however, was probably the restoration of the links with British overseas territories that had weakened since they were first established by Sir Joseph Banks. Hooker guided the development of the rubber plantations of Malaysia and India and the introduction of Liberian coffee to Sri Lanka.

JEAN-BAPTISTE LAMARCK AND THE ROYAL GARDEN, PARIS

In 1778 a former Paris bank clerk published *Flore Française* (French flora) in three volumes. The work was an immediate success and, with the support and help of France's leading natural scientist, Georges-Louis Leclerc, comte de Buffon (1707–88), its author was appointed an assistant botanist at the Jardin du Roi, the royal botanical garden in Paris. The bank clerk turned botanist was Jean-Baptiste Lamarck (1744–1829), or to give him his full name and title Jean-Baptiste Pierre Antoine de Monet, chevalier de Lamarck. Despite his grand-sounding name, Lamarck was poor for most of his life, and he died in poverty.

He was born on August 1, 1744, in Bazentin-le Petit, a village in the Picardy region of northeastern France. The Lamarck family was aristocratic but far from wealthy, and Jean-Baptiste was the youngest of 11 children. His father was a soldier, and several of Jean-Baptiste's brothers had joined the army, but Lamarck's father wished him to enter the church and sent him to a Jesuit seminary at Amiens. After his father's death in 1760, however, Lamarck left the seminary, acquired a horse and a letter of introduction to a colonel, and rode to Germany, where the French army was fighting. He joined the army on the eve of a battle, and the next day he showed such heroism and leadership under fire that he was made an officer on the spot. The war ended in 1763, and Lamarck's regiment was sent to Monaco. While there, one of his comrades lifted him by the head during a bout of horseplay, causing a neck injury that necessitated surgery and a long period of recuperation. It was during this period that he became interested in botany. The army awarded him a commission and pension, but the pension was too small to live on so Lamarck left the army and traveled to Paris to study medicine, taking a job as a bank clerk to support himself. He abandoned his medical studies after four years and turned to botany, studying under the botanist Bernard de Jussieu (1699–1777). Lamarck's *Flore Française* was the product of the 10 years he spent studying French plants.

Lamarck was elected to the French Academy of Sciences in 1779, and in 1781 he was appointed royal botanist. His position at the Jardin du Roi allowed him to travel, and he collected plants for the royal garden. He also collected mineral specimens and other items for French museums.

The Jardin du Roi had been established in 1626 as a physic garden, and it opened to the public in 1640. Lamarck became a professor of botany there in 1788 and was placed in charge of the herbarium. During the French Revolution, the name became an embarrassment, and in 1790 Lamarck changed it to Jardin des Plantes (botanical garden). In 1793, with Lamarck's warm approval, the institution was reorganized once more, becoming part of the Muséum national d'Histoire naturelle. The museum was to be run by 12 professors, and Lamarck was appointed to what his colleagues considered the least important position, as professor of insects and worms—his official title was professeur d'histoire naturelle des insectes et des vers. Until that time Lamarck had not studied zoology, but now he dedicated himself to it and coined the word *invertebrates* to describe the animals that had become his specialty. He went on to write several books on zoology. In addition to the *Flore Française,* he wrote a three-volume *Dictionnaire de Botanique* (Dictionary of botany), published between 1783 and 1789; *Illustrations des Genres* (Illustrations of the genera), published between 1791 and 1800; and *Histoire naturelle des végétaux* (Natural history of vegetables), published in 1803.

Lamarck married Marie Anne Rosalie Delaporte in 1778, and they had five children. Marie died in 1792, and the following year Lamarck married Charlotte Reverdy. She died in 1797. In 1798 Lamarck married Julie Mallet, who died in 1819. In about 1818 Lamarck's eyesight began to fail, and after some years he became totally blind. His daughters were devoted to him and cared for him, but the family was desperately poor. He died on December 28, 1829, and received a pauper's funeral and was buried in a rented grave. His body was removed in about 1835, and its present location is unknown. His books and the contents of his home were sold.

Today the Jardin des Plantes is one of the seven departments of the Muséum national d'Histoire naturelle. It occupies 70 acres (28 ha) beside the River Seine in Paris and is France's principal botanical garden, with a botany school that trains botanists. The garden has approximately 4,500 plant species arranged taxonomically in a 2.5-acre (1-ha) plot, and 7.5 acres (3 ha) are devoted to ornamental plants. There is a display of alpine plants with about 3,000 species and a rose garden.

JOSÉ MUTIS AND THE BOGOTÁ BOTANICAL GARDEN

The Jardín Botánico José Celestino Mutis in Bogotá, Colombia, is a research and scientific center and the country's largest botanic garden. It occupies 20 acres (8 ha) and specializes in Andean species, with greenhouses containing plants from every region of the country, including 5,000 species of orchids and the world's largest example of the giant Amazonian water lily (*Victoria regia*).

The garden was established in 1955 by the Jesuit priest Enrique Pérez Arbeláez and is named after one of the most important Spanish botanists of the 18th century, José Celestino Bruno Mutis (1732–1808). In 1764, Mutis asked for financial support to create a botanical garden in Bogotá, but no funds were available and he was refused. However, Mutis did establish a botanical garden in the city of Mariquita, built up a large botanical library, and assembled a herbarium with more than 24,000 plants. He also possessed about 5,000 plant drawings made by his pupils. His book *Flora de Bogotá o de Nueva Granada* (Flora of Bogotá or New Granada) contained more than 6,000 illustrations. He was able to send the manuscript home to Spain, but the government could not afford to publish the work.

Mutis made a special study of the genus *Cinchona*, which contains 23 species of trees and shrubs that occur naturally from Bolivia northward to Colombia and Venezuela, with one found in Costa Rica. A number of chemical substances are extracted from them, including quinine, which is still used to treat malaria. In Mutis's day, physicians believed cinchona would cure any number of illnesses.

Mutis described the geographical distribution of cinchonas and listed all the species and varieties, with their therapeutic uses. He wrote several articles on cinchona and quinine, but although he sent the manuscript of his most important book on the subject, *El arcano de la quina* (The mystery of cinchona), to Madrid, it arrived at a time when Spain was at war with France, and it was not published. Part of the work was finally published in 1828, and the scientific part was discovered later in a shed in the Madrid botanical gardens. That text was published in 1867 and the accompanying tables in 1870.

José Celestino Bruno Mutis y Bosio was born in Cádiz, Spain, on April 6, 1732. He studied medicine, first at the Cádiz College of Surgery and then at the University of Seville, from which he graduated in 1755 and received his doctorate in medicine in 1757. He taught

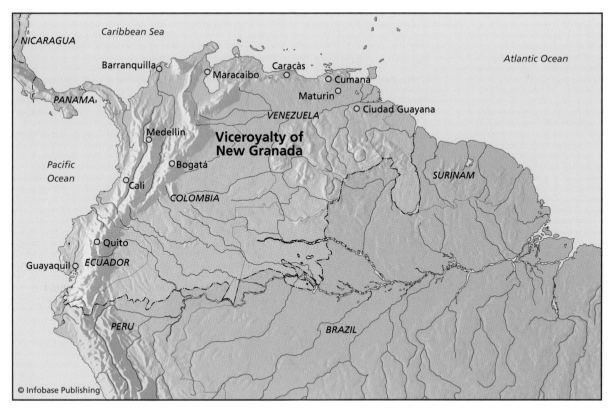

The Viceroyalty of New Granada (Virreinato de la Nueva Granada) was the name given in 1717 to the Spanish colonies in the northern part of South America. The region approximately covered the modern countries of Panama, Colombia, Ecuador, and Venezuela, as well as part of Peru, Brazil, Guyana, and Nicaragua.

anatomy in Madrid from 1757 until 1760, at the same time studying botany at what was then the Migas Calientes Botanical Gardens in Madrid (now the Real Jardín Botánico de Madrid). Mutis also studied mathematics and astronomy. On September 7, 1760, Mutis sailed from Spain to become the principal physician of the viceroy of New Granada, arriving in Bogotá on February 24, 1761. In 1772 Mutis was ordained a priest. New Granada was the name given to the Spanish colonies in northern South America. The map on page 77 shows its extent.

Soon after his arrival, Mutis began his botanical investigations, and in 1763 he proposed to the Spanish king that he undertake an expedition to study the flora and fauna of New Granada. It took the king until 1783 to authorize this, and that was when the royal botanical expedition that Mutis led began. It continued for 25 years and covered approximately 3,000 square miles (8,000 km²). Mutis collected specimens, labeling them and writing detailed descriptions of the places he found them. He sent all of his vast collection of material back to Spain in 105 boxes, where the plants, manuscripts, and drawings were stored in a shed in the botanical gardens. In 1801 the German explorer Alexander von Humboldt (see "Alexander von Humboldt and the Plants of South America" on pages 110–113) stayed with Mutis for two months and was greatly impressed by Mutis's botanical collection. Mutis died in Bogotá on September 11, 1808.

Naming Plants

A plant that grows naturally across a wide geographical region has a name in each of the languages of that region. In many countries it is likely to have more than one local name, and the same name may refer to different plants in different places. Viper's bugloss, for instance, is known as blueweed in some places. Jewelweed is also called balsam and touch-me-not, but there are about 20 plants with balsam in their names, including balsam apple, balsam bog, Canada balsam, Copaiba balsam, balsam fig, balsam fir, gurjun balsam, Indian balsam, Mecca balsam, balsam of Peru, and Tolu balsam—and they are all different. Cowslip refers to different, unrelated plants in Britain, North America (where there are 14), and South Africa.

Plant names are confusing, and this chapter recounts the story of how the confusion was finally resolved. Today, a botanist in one part of the world is able to discuss any plant with a botanist in a distant country who speaks a different language, and there is no risk of ambiguity because all botanists agree to give each plant its own unique name. For example, Mecca balsam is *Commiphora opobalsamum* and the 14 North American cowslips are 13 species of *Dodecatheon,* also called shooting star, and Virginian cowslip is *Mertensia virginica,* also known as Virginian bluebells. Local names survive, and long may they continue to do so, but botanists have no need or use for them.

Agreeing on a single system for naming plants also involved agreeing on a system for classifying them, and this chapter explains how that system came into being and how it works. Many naturalists had attempted to classify plants (see "John Ray and His Encyclopedia of Plant Life" on pages 16–18), but this chapter confines itself to describing the development of the system that is used today. It ends by telling of the birth of a new branch of the plant sciences: plant geography.

JOSEPH PITTON DE TOURNEFORT AND THE GROUPING OF PLANTS

Its *flower* is the most obvious feature of most flowering plants. Many flowers have brightly colored petals. These attract pollinating animals. Plants such as grasses that are pollinated by wind have no need of petals, so their flowers are usually smaller and drab in color, but they are nevertheless prominent and clearly visible. However it is pollinated, the flower is the plant's reproductive structure, containing either male organs or female organs or both. The following illustration shows an idealized flower, with petals and both male and female reproductive organs. A flower of this type is said to be *perfect.* Flowers with only *stamens*—male reproductive organs—are said to be *staminate,* and flowers with only *carpels*—female reproductive organs—are *carpellate.* Staminate and carpellate flowers are described as *imperfect.* In addition, the flower in the illustration has both petals and *sepals*—leaflike structures that together formed the *calyx* that enclosed and protected the flower bud. The possession of petals and sepals makes the flower complete. An incomplete flower lacks either petals or sepals or both.

When botanists set about classifying plants and teaching their students techniques for identifying them, it is not surprising that they first concentrated on the flowers. In 1694 the French botanist Joseph Pitton de Tournefort (1656–1708) published a guide to plant recognition based on flower structure that prepared the way for Linnaeus to develop his classification system (see "Carolus Linnaeus and the Binomial System" on pages 83–86). The guide was entitled (in the original 17th-century French) *Elémens de botanique, ou Méthode pour connoître les Plantes* (Elements of botany, or method for recognizing plants). Tournefort wrote in French, and in 1700 the work was translated into Latin with the title *Institutiones rei herbariae,* with a second edition published in 1719. The French edition contained 451

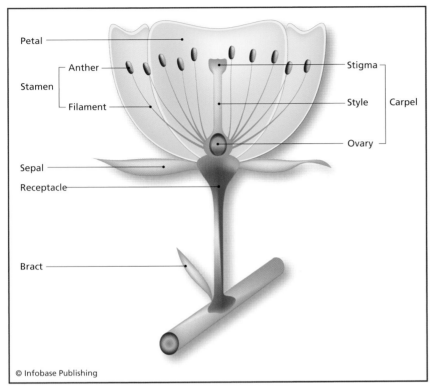

© Infobase Publishing

The flower is the reproductive apparatus of every flowering plant. This flower carries both male (stamens) and female (carpel) reproductive organs, enclosed in petals. Flowers of some plants hold either, but not both, male or female organs, and some flowers (e.g., of grasses) have no petals.

plates of illustrations and the Latin translation had 476. In 1703 a supplement was published with an additional 13 plates. The work had three volumes, the first containing the text and the remaining two containing the illustrations.

Tournefort included a description of *fungi* and classified *lichens* as a group distinct from plants. Fungi were long thought to be plants, but are nowadays recognized as belonging in a separate taxonomic kingdom. A lichen is a composite organism consisting of a fungus together with an *alga*—a simple plantlike organism that is not differentiated into roots, stems, and leaves—or a *cyanobacterium*—a bacterium that performs photosynthesis. Tournefort based his key to plant identification on the form of the *corolla*—the structure formed by all the petals. In other words, he taught his students to recognize plants by the outward appearance of their flowers. On this basis, he grouped plants into 22 classes, but then used other characteristics such as the shape of the fruit and seed to divide the classes into sections and genera. His method was easy to use, because flowers vary greatly. More important, however, Tournefort was the first botanist to make a clear distinction between species and genus. His *Elémens de botanique* described approximately 7,000 plants and he was able to arrange these into about 700 genera. Linnaeus used many of the generic and specific names Tournefort introduced. At higher taxonomic levels, however, Tournefort continued the old practice of designating plants as trees, shrubs, or herbs.

Simple though Tournefort's system was, it had two major drawbacks. The first and most obvious was that only flowering plants can have a corolla, and even then wind-pollinated plants have flowers lacking petals. Coniferous plants such as pines, spruces, firs, redwoods, and junipers produce male and female cones. These are colored in some species, but they are not true flowers, and they have no petals. So entire groups of plants remained outside the system. The second drawback was more subtle, but no less serious. Tournefort's system was artificial. Related species often have very similar flowers but not always, and two species with similar flowers may be related only distantly or not at all. The structure of their flowers has no bearing on the relationship between species.

Joseph Pitton de Tournefort was born on June 5, 1656, in Aix-en-Provence in southern France. He commenced his education at the Jesuit seminary in Aix, where he studied theology with the intention of entering the church, but after his father's death he turned to botany and spent two years collecting plants before studying medicine at the universities of Montpellier and Barcelona. In 1683 he was appointed director and professor of botany at the Jardin du Roi in Paris, a position that allowed him to travel extensively in southern Europe on behalf of the king, who instructed him to check the accuracy of maps and to observe conditions in the countries he visited, traveling as much as possible overland rather than by sea. Between April 1701 and June 1702, Tournefort passed through the Greek islands on his way to visit Constantinople (modern Istanbul) and the lands bordering the Black Sea, including Armenia and Georgia and extending to the Persian (Iranian) frontier. He collected plants on all his travels, and during his journey to the east he collected specimens of 1,356 species. His account of his travels was published posthumously in 1717 as *Relation d'un voyage du Levant* (Account of a journey in the Levant). He was elected a member of the Academy of Sciences in 1692. Tournefort qualified as a doctor of medicine in 1698, and in 1702 he became professor of medicine at the Collège de France. Tournefort died in Paris on December 28, 1708, in an accident with a carriage on the street that is now called the rue de Tournefort.

CAROLUS LINNAEUS AND THE BINOMIAL SYSTEM

A common plant is likely to have many names, but to avoid confusion it will have only one scientific name. The scientific name will be in Latin and it will have two parts, the first being the name of the genus and the second of the species. This use of two names to identify a species is called the *binomial system* (see the following sidebar "How Plants Are Classified"). Caspar Bauhin and his brother Jean (see "The Bauhin Family" on pages 42–43), working in the late 16th century, were the first botanists to use the binomial system, but the scientist who developed it and set it within a broader taxonomic concept was the Swedish botanist Carolus Linnaeus (1707–78). In 1735 Linnaeus published the first edition of *Systema naturae per regna tria naturae, secundum classes, ordines, genera, species, cum characteribus, differentiis, synonymis, locis* (System of nature through the three kingdoms of nature, according to classes, orders, genera, species, with characters, differences, synonyms, places), the book describing his taxonomic system. It had only 11 pages, but by the time *Systema naturae* reached its 13th and final edition, published in 1770, it had grown to about 3,000 pages and listed 4,400 species of animals and 7,700 species of plants. In 1737 Linnaeus published his book *Genera plantarum* (Plant genera) and in 1753 he published *Species plantarum* (Plant species). In *Species plantarum* Linnaeus described every species of plant known at the time and arranged the species within the taxonomic scheme he had devised. These two works are to this day the scientific starting point for the classification of all flowering plants and ferns. Linnaeus brought together the botanical material in *Systema naturae* with the material in *Genera plantarum* and *Species plantarum* in his

Carolus Linnaeus (1707–78), in his early 30s, wearing Lapp dress and holding a plant. He acquired the costume during his 1732 expedition to Lapland and wore it for this famous portrait. *(Mansell/Time Life Pictures/Getty Images)*

1779 work *Systema plantarum* (Botanical system). In total Linnaeus wrote about 180 books.

Carl Linnaeus was born on May 23, 1707, on a farm at Råshult, in Småland, southern Sweden. Linnaeus wrote in Latin, and Carolus is the Latin version of his name, which he used in his professional life. His father, Nils Ingemarsson Linnaeus, was a clergyman who, as was

HOW PLANTS ARE CLASSIFIED

Most plants, though not all, have a name in the language of every country in which they occur naturally. In many cases they also have several local names. In addition, every plant known to science has an "official" or scientific name. This name is in Latin, by convention it is written in italic characters, and it has two parts. The first, written with an initial capital letter, is the name of the genus, and the second, with an initial lower-case letter, is the name of the species. Subspecies are given an italicized name following the specific name. With some plants there is more than one variety of a species or subspecies, generating an additional italicized name preceded by the abbreviation "var." for "variety."

Scientific plant names are devised according to a set of rules contained in the International Code of Botanical Nomenclature. The person proposing a name must submit it, together with an explanation of the name and a specimen of the plant (usually a herbarium specimen), to the International Committee on Botanical Nomenclature, which is the body that administers the code. If the committee accepts the name, the person who devised it may be identified by initials following the full name. For example, the common daisy, which was named by Linnaeus, is *Bellis perennis* L., and the California redwood is *Sequoia sempervirens* D. Don, for David Don (1799 or 1800–1841).

Higher taxonomic categories exist above the level of genus. Genera are grouped into families, families into orders, orders into classes, and classes into phyla. Above the level of phylum, all plants belong to the kingdom Plantae, and the Plantae belongs to the domain Eukarya. The full classification of *Bellis perennis* is as follows:

Domain Eukarya (or Eucarya) (Organisms with eukaryotic cells)
Kingdom Plantae (or Metaphyta) (All plants)
Phylum (or division) Anthophyta (flowering plants)
Class Eudicots (dicotyledons)
Order Asterales (plants related to sunflowers)
Family Asteraceae (sunflower family)
Genus *Bellis*
Species *perennis*
Name *Bellis perennis* L.

Where necessary, taxonomists add intermediate ranks between the main ranks, such as subkingdom, subphylum, superphylum, subclass, superclass, suborder, superorder, subfamily, superfamily, and subspecies.

the custom at the time in Småland, had originally been known by his patronymic name: Ingemarsson means son of Ingemar. But when Nils enrolled at university, the authorities needed to know his surname, so he gave himself the name Linnaeus, after an old and much-loved lime tree (*linn* in the Småland dialect) on land the family owned. Nils was a keen gardener and familiar with the wild flowers growing in the area, and by the time Carl was four his father was teaching him about plants and encouraging him to learn their names. Carl's passion for naming plants came to shape his adult life.

Linnaeus began his education with a tutor and then attended a school in the nearby town of Växjö before entering the University of Lund in 1727 to study medicine. The following year he transferred to the University of Uppsala, where there was a botanical garden in which he spent as much time as he could. Linnaeus became friendly with Olof Celsius (1701–44), the dean of Uppsala Cathedral, and Celsius, recognizing the young man's enthusiasm for plants, introduced him to Olof Rudbeck (1660–1740), the garden's director. Rudbeck was approaching retirement and he invited Linnaeus—a first-year student—to deliver botany lectures. It was at about this time that Linnaeus began to feel his way toward devising a system for classifying plants on the basis of their flower structure along the lines Tournefort had pioneered.

In 1732 Linnaeus spent four months exploring Lapland, then a vast wilderness extending across northern Sweden, Finland, and Russia. He walked to the shore of the Arctic Ocean, collecting plants all the way. The following year Linnaeus became engaged to Sara Elisabeth Moraeus, the daughter of Johan Moraeus, a physician who advised him to complete his medical studies in the Netherlands. Linnaeus did so, qualifying in 1735 at the University of Harderwijk, after which he spent a short time in England before returning to the Netherlands. His meeting later that year with George Clifford (see "The Rise of the Herbarium" on pages 60–62) led to Linnaeus compiling the *Hortus Cliffortianus,* a book describing Clifford's plant collection.

Linnaeus began practicing medicine in Stockholm in 1738. His practice was successful and allowed him time to continue his botanical studies. He was a founding member of the Royal Swedish Academy of Sciences in 1739. He was appointed professor of medicine at the University of Uppsala in 1741, but within a year he exchanged that post to become professor of botany. King Adolf Fredrik made

Linnaeus a Knight of the Polar Star in 1758 and the Privy Council confirmed his ennoblement in 1761. From that time Linnaeus took the name Carl von Linné.

Von Linné's health began to fail in 1772, and in 1774 he suffered a stroke from which he made a partial recovery. The illustration on page 83, from an engraving made in 1775, shows him in his early 30s, wearing Lapp dress. A second stroke in 1776 paralyzed his right side. He died on January 22, 1778, during a ceremony in Uppsala Cathedral, where he is buried.

AUGUSTIN DE CANDOLLE AND NATURAL CLASSIFICATION

Linnaeus developed the approach to classification that Tournefort had used, but he did not depart from its principles. Under the Linnaean system, plants were grouped together on the basis of similarities in their flowers—the number and arrangement of stamens. His system was easy to learn, so students liked it, but it was unnatural, and in later years much of it had to be dismantled and the groups of species rearranged—requiring former students to painfully unlearn the arrangements they had memorized.

An alternative approach was possible. In 1790 the German poet Johann Wolfgang von Goethe (1749–1832) published *Versuch, die Metamorphose der Pflanzen zu eklären* (Attempt to explain the metamorphosis of plants). Goethe believed that the leaf is the basic unit of every plant and that different forms of plants arise through the expansion and contraction of leaves and of organs derived from the modification of leaves. All plant species were, therefore, variations on a single general plan. This idea influenced the Swiss botanist Augustin de Candolle (1778–1841). Candolle was also influenced by the French botanist Antoine-Laurent de Jussieu (1748–1836). In 1788–89, Jussieu, professor of botany at the Jardin des Plantes (1793–1826), published his classification system in his book *Genera plantarum* (Plant genera). Jussieu had classified plants by first dividing them into three divisions according to whether they had no cotyledons in their embryos (acotyledons), one cotyledon (monocotyledons), or two cotyledons (dicotyledons). He then broke the divisions into 100 families on the basis of other natural features.

Taxonomy was Candolle's passion; indeed, it was he who coined the term in his 1813 book *Théorie élémentaire de la botanique* (Elementary theory of botany), which was the work in which he set out his system of classification—or taxonomy. Candolle based this on the work of Jussieu and the French zoologist Georges Cuvier (1769–1832). Candolle proposed that similarities in the symmetries of their sexual parts revealed whether plants were related, but that these similarities could be obscured by parts fusing together, or degenerating, or by being lost entirely. These modifications could make plants that were truly related appear to be unrelated. To overcome such difficulties, Candolle introduced the concept of *homology*—the existence in two species of organs or structures that appear different but that are descended from a common ancestor.

Candolle was also the first botanist to recognize that plants compete with each other—he called it being "at war one with another"—for resources. Charles Darwin (see "Charles Darwin and Evolution by Means of Natural Selection" on pages 151–154) studied Candolle's plant taxonomy while he was a student at the University of Edinburgh, and the description of plants "at war" was one of the ideas that pushed him toward developing his own evolutionary theory.

Augustin Pyramus de Candolle was born in Geneva, Switzerland, on February 4, 1778. He studied medicine at the Geneva Academy for two years before moving to Paris in 1796, where he added natural sciences to his medical studies. While in Paris, Candolle met many leading scientists, including Georges Cuvier and Jean-Baptiste Lamarck (see "Jean-Baptiste Lamarck and the Royal Garden, Paris" on pages 74–76). In 1802 he acted as Cuvier's deputy at the Collège de France, and he edited the third edition of *Flore française* on behalf of Lamarck. He received his degree of doctor of medicine in 1804 from the University of Paris, but his acquaintance with Cuvier and Lamarck led Candolle to abandon medicine and concentrate wholly on botany. At the request of the French government, he carried out a botanical and agricultural survey of the whole of France. It took him six years, traveling each summer, and the results were published in 1813. In 1808 Candolle left Paris to become professor of botany at the College of Medicine in Montpellier. He left Montpellier in 1816 to become professor of natural history and director of the botanical garden at the Geneva Academy. He remained in these positions until he retired in 1835, when his son Alphonse Louis Pierre Pyramus de Candolle (1806–93) succeeded him and continued his work.

In 1824 Candolle commenced his most important work, *Prodromus systematis regni vegetabilis* (Introduction to a natural classification of the vegetable kingdom). He intended this 17-volume treatise to cover the taxonomy, ecology, and geography of all known seed plants. By 1839 he had completed seven of the volumes, but then ill health compelled him to abandon work on the project. He died in Geneva on September 9, 1841. Alphonse continued the work, producing the remaining 10 volumes, one of them in collaboration with his own son, Anne Casimir Pyrame de Candolle (1836–1918).

ADOLF ENGLER AND THE VEGETATION OF THE WORLD

In 1924 the German botanist Adolf Engler (1844–1930) published *Syllabus der Pflanzenfamilien* (Summary of plant families). In this work, Engler set out his own system of plant taxonomy. Many herbaria, field guides, and floras still use Engler's classification, and the 12th edition of this two-volume work, edited by H. Melchior and E. Werdermann, was published in 1964. In addition to his taxonomic system, Engler collaborated with the German botanist Karl Anton Eugen Prantl (1849–93) in editing a 23-volume *Die Natürlichen Pflanzenfamilien* (The natural plant families), with contributions from many other botanists and more than 33,000 botanical drawings in about 6,000 plates drawn by Joseph Pohl (1864–1939). In 1986 the International Association for Plant Taxonomy instituted the Engler Medal in his honor, to be awarded to scientists who have made outstanding contributions to plant taxonomy.

Engler was also a pioneer in the new science of *phytogeography*—the study of the geographical distribution of plants. From 1889 to 1921, he was director of the Berlin Botanische Zentralstelle—the Botanical Garden and Museum. One of the main purposes of the Berlin garden was to celebrate German colonial achievements. It listed and where feasible cultivated plants from all the German colonies, arranged in beds geographically, and conducted experiments in acclimatization that might aid the movement of useful plants from one part of the world to another. The decision was made in the 1890s to transfer the garden to a new, larger site in the district of

Dahlem, and Engler supervised the move, which began in 1897 and was completed in 1910. This was a major operation, and it was one that allowed Engler to arrange the layout in a way that would display natural formations of plants along the lines August Grisebach (1814–79) had described in 1883 (see "August Grisebach and Floral Provinces" on pages 122–124). The Berlin-Dahlem Botanical Garden is now linked to the Free University of Berlin and is one of the world's most important botanical gardens, displaying approximately 22,000 plant species in its 106 acres (43 ha).

Between 1896 and 1923, in collaboration with the German botanist and biogeographer Carl Georg Oscar Drude (1852–1933), Engler published *Die Vegetation der Erde* (The vegetation of the Earth). He published *Die Pflanzenwelt Ost-Afrikas und der Nachbargebiete* (The plant world of East Africa and neighboring regions) in 1895 and *Die Pflanzenwelt Afrikas* (The plant world of Africa) in 1910. In 1880 he founded the *Botanischer Jahrbücher für Systematik, Pflanzengeschichte und Pflanzengeographie* (Botanical yearbooks for taxonomy, plant evolution, and plant geography), and he edited them from 1880 until 1930.

Heinrich Gustav Adolf Engler was born on March 25, 1844, in what was then Sagan, Prussia (now Żagań, Poland). He studied at the University of Breslau, Germany (now Wrocław, Poland), where he received his Ph.D. in 1866. Engler then worked as a schoolteacher until 1871, when he was appointed a lecturer and curator of the herbarium specimens at the Munich Botanical Institute. In 1878 Engler became professor of systematic botany at the University of Kiel. He remained at Kiel until 1884, when he moved to the University of Breslau as professor of systematic botany and director of the botanical gardens. In 1889 Engler became professor of botany and director of the botanical gardens at the Berlin Botanical Gardens and Museum, where he remained until he retired in 1921. Engler died in Berlin on October 10, 1930.

The Plant Hunters

Although most of them traveled extensively, the great botanists such as Linnaeus, Tournefort, and Engler could not explore every corner of the world. Most of the plants that they studied, classified, preserved in their herbaria, and cultivated whenever possible were sent to them by specialist plant collectors. Linnaeus recruited his former students to this task and called them "the Apostles." There was also a commercial aspect to plant collecting. Some plants had agricultural potential. Many more could ornament private parks and gardens, and landowners were willing to pay handsomely for the latest exotic plant to arrive from some remote and inaccessible region. Once they arrived, specimens were cultivated in selected sites where they were tended by experienced botanists and growers until they were acclimatized to their new environment. Once acclimatized, they found their way first to the gardens of the wealthy and later, as they became more plentiful and the price fell, to keen amateur gardeners of more modest means.

The great botanical gardens continue to send explorers out into the world in search of new plant species. Nowadays, of course, they work in collaboration with botanists in the countries they visit and must first obtain the permission of the host government. The specimens they collect and remove are for scientific research only. Quite apart from the right of all governments to regulate the exploitation of natural resources within their own borders, the risk of importing plant pests and diseases severely restricts the international move-ment of plant materials.

This chapter tells the story of a few of the great collectors, the places they visited, and the plants they sent home.

RHODODENDRONS, PRIMULAS, AND FRANK KINGDON-WARD

Rhododendrons have long been popular with gardeners. There are approximately 850 species in the genus *Rhododendron*. At least 26 species occur naturally in North America. The following illustra-tion shows one of these, Catawba rhododendron (*R. catawbiense*). In China there are about 650 species and large numbers grow in the mountains of Nepal, Bhutan, and Sikkim. The mountains of New Guinea have about 150 species that are found nowhere else.

With so many species, rhododendrons come in a variety of forms. Some are evergreen, some deciduous, most are shrubs or trees but some are climbers, and all of them are poisonous—with some species, honey made from the flowers is also poisonous. They grow in moun-tainous regions, which means that despite being tropical plants they prefer cool temperatures and high rainfall, requirements that make them ideally suited for cultivation in temperate regions.

Rhododendrons and azaleas, which are hybrid rhododendrons, produce large, showy flowers, and Japanese growers were produc-ing hybrids in the 17th century. Wealthy European landowners also began cultivating them in the 17th century. The first rhododendron in Britain was introduced in 1656. This was the alpine rose (*Rhodo-dendron hirsutum*), native to the Alps, which had been named by Clusius (see "Carolus Clusius, the Leiden Botanical Garden, and the Tulip" on pages 64–66). Another rhododendron called the alpine rose, *R. ferrugineum,* was introduced to Britain in 1752. Several North American species were also taken to Britain in the 18th century. Rhododendrons are now cultivated wherever the soil is suitable. At least 500 species have been introduced in Britain and *R. ponticum*, a Mediterranean species imported from Gibraltar, has escaped cultiva-tion and become a highly invasive weed.

The demand for these ornamental plants made it profitable for botanical expeditions to collect them. Sir Joseph Hooker, director

A Catawba rhododendron (*Rhododendron catawbiense*) growing in the Blue Ridge Mountains of North Carolina. This species is native to the eastern United States, growing mainly in the Appalachian Mountains, and is named after the Catawba tribe of Native Americans. *(Adam Jones/Visuals Unlimited)*

of Kew Gardens, returned to Britain with 45 new species he had found in Sikkim. In 1854 and 1855, Robert Fortune (see "Robert Fortune, Collecting in Northern China" on pages 101–105) discovered *R. ovatum* and *R. fortunei.* These became the source of a series of hybrids that are widely grown in the United States. Between them, George Forrest and Ernest Wilson returned hundreds of species to Europe (see "George Forrest, Collecting in Yunnan" on pages 98–101 and "Ernest Wilson, Collecting in China and Japan" on pages 106–108).

Frank Kingdon-Ward (1885–1958), a botanist and the son of Harry Marshall Ward, a professor of botany, devoted his life to hunting plants in Asia, and his favorite plants were rhododendrons, primulas, and gentians. He made a total of 25 expeditions and discovered 10 species of rhododendrons and two previously unknown primulas in Assam, as well as many other plants. *R. wardii,* one of the rhododendrons he found, has yellow flowers, and the primulas included the giant primrose (*Primula florindae*). He wrote 25 books describing his travels and his many adventures. The illustration at right shows him in later life.

Francis Kingdon Ward (at first there was no hyphen and his middle name was his mother's maiden name) was born on November 6, 1885, in Manchester, where his father was professor of botany at Owen College. In 1895 Frank's father became professor of botany at the University of Cambridge. Kingdon-Ward was educated at St. Paul's School in London, and in 1904 he entered Christ's College, Cambridge. His father died in 1906, however, compelling Frank to cut short his education and seek paid employment, but he obtained a degree after two years and agreed to complete his third year later. A friend found him a job teaching in a school in Singapore. He spent two years there before joining a zoological expedition looking for new species along the Yangtze River. He sent a small number of plants back to Cambridge, as well as discovering two previously unknown species of shrew and one new mouse. In 1910 he published an account of his experiences in his book *On the Road to Tibet,* and the following year he was elected a fellow of the Royal Geographical Society and commissioned by a seed company to explore Tibet and Yunnan Province of China looking for hardy plants that would grow in British gardens. Kingdon-Ward collected about 200 specimens, including 22 plants that were new to science, and, as he did from all of his expeditions, he sent a proportion of his finds to the herbarium at Kew Gardens. One of his discoveries was the blue poppy (*Meconopsis speciosa*). Unfortunately, this would not grow in Britain, where the cultivated Himalayan blue poppy is *M. betonicifolia,* a different species, but it gave him the title for his next book, published in 1913—*The Land of the Blue Poppy.* His next expedition took him back to Yunnan and Tibet. In 1914 Kingdon-Ward was in Myanmar (Burma) when World War I began. He enlisted in the army, returning to Myanmar as soon as the war ended.

Kingdon-Ward returned to both Myanmar and Tibet several times over the following years. He visited the United States in 1939, but returned to Britain in time to volunteer for service in World War II. Too old for active service, he was initially employed as a censor

Frank Kingdon-Ward collected plants in eastern Asia, specializing in rhododendrons and primulas. This portrait was painted by Miss E. M. Gregson. *(Royal Geographic Society, London)*

and translator of Chinese and other Asian languages. Later, posing as a botanist working for the government, he was sent to Myanmar to search for a route between India and China. He ended the war as an instructor, training airmen in jungle survival. Toward the end of 1947 he and his second wife embarked on an expedition to Manipur, India, collecting herbarium specimens of about 1,000 plant species. There were further expeditions to the border between Assam and Tibet on behalf of the New York Botanical Garden and the Royal Horticultural Society, where the couple collected 37 species of rhododendron and nearly 100 other species. His last expedition, in 1956, was again to Myanmar.

Kingdon-Ward married twice, to Florinda Norman-Thompson in 1923 and to Jean Macklin in 1947; his first marriage ending in divorce. In the course of his travels he suffered many injuries and caught malaria, which recurred at intervals. He suffered a stroke in 1958, from which he died on April 8 without regaining consciousness.

DAVID DOUGLAS IN NORTH AMERICA AND HAWAII

The Douglas fir (*Pseudotsuga menziesii*) is a magnificent tree that occurs naturally in western North America from British Columbia southward to northern Mexico. It can grow up to 330 feet (100 m) tall and is one of North America's most valuable sources of timber. The tree can occur in dense stands in forests, but it also makes a handsome ornamental tree wherever there is space for it. The following illustration shows a fine specimen growing in the open, where it can attain its full size.

The "Douglas" in its name refers to the Scottish botanical explorer David Douglas (1799–1834), who also gave his name to hundreds of other western plants. He described 50 species of trees, including the sugar pine (*Pinus lambertiana*), and more than 100 other plants, and he was responsible for introducing more than 250 species to Britain. Douglas was the first European who was not a fur trader to explore the interior of British Columbia.

The tree's specific name refers to another Scottish botanist and naval surgeon Archibald Menzies (1754–1842). It was Menzies who brought the Chile pine or monkey puzzle tree (*Araucaria araucana*) to Europe. While dining with the governor of Chile, seeds from this tree were served as dessert. Menzies managed to conceal a few in his pocket and later germinated them on board the ship on which he was serving, returning to England with five healthy plants.

David Douglas was born on June 25, 1799, in Scone, Perthshire, the son of a stonemason. He attended a school in nearby Kinnoull, but left at the age of 10. When he was 11, he was apprenticed to William Beattie, the head gardener at Scone Palace. Scone Palace was formerly where the kings of Scots were crowned, but in Douglas's time it was the home of the earl of Mansfield. Later, Douglas moved to the Valleyfield estate of Sir Robert Preston, where he was allowed access to the library and rose to the position of under-gardener. His next move was to Glasgow Botanic Gardens in 1820. There, William Hooker, then professor of botany at Glasgow University (see "Sir William Hooker, the First Official Director" on pages 72–74), befriended him, taking him on botanical excursions and teaching him how to prepare herbarium specimens. Hooker recommended Douglas as a plant collector to the Royal Horticultural Society in London, and in 1823 the society sent him to collect plants in New York and eastern Canada. This trip was successful, and the following year the society sent Douglas to collect specimens in the Columbia River region of northwestern Canada. He sailed from London on the Hudson's Bay Company ship *William and Ann* in July 1824, rounded Cape Horn, and reached Fort Vancouver in April 1825.

The Douglas fir (*Pseudotsuga menziesii*) is a magnificent tree that can grow up to 330 feet (100 m) tall. *(Michael P. Gadomski/Photo Researchers, Inc.)*

Douglas began collecting at once and spent the remainder of 1825 and all of 1826 exploring the mountains of northeastern Oregon. In March 1827, accompanied by a small party, Douglas headed for Hudson Bay on his way back to England. Following the Hudson's Bay Company's route known as the York Factory Express, they covered 2,600 miles (4,180 km), traveling by boat up the Columbia River, crossed the Rocky Mountains, and on July 28 they arrived at York Factory on the shore of the bay, where they embarked for England. The following map shows the route of the York Factory Express.

In October 1829 Douglas left England once more, arriving at the Columbia River the following June, and in October 1830 he entered

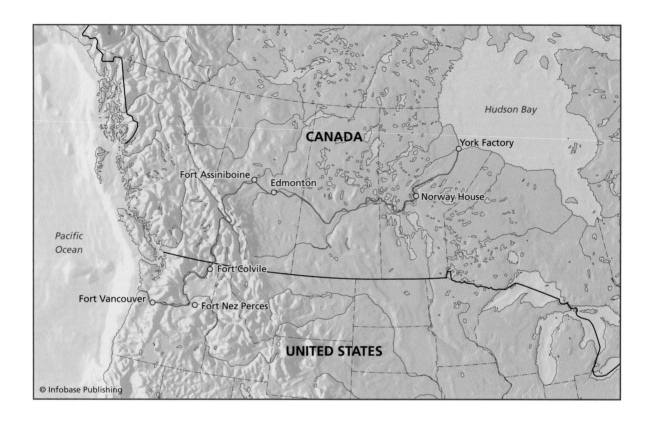

© Infobase Publishing

The York Factory Express, so called because it was used to transport company documents as well as furs and other goods, was a route between Fort Vancouver and York Factory that was operated by the Hudson's Bay Company in the first half of the 19th century.

California, where he remained until August 1833, collecting about 500 species of plants and also many mosses, which were of special interest to William Hooker. Douglas spent a short time in Oregon, and in October he sailed for Hawaii, which he was in the habit of visiting every winter collecting plants. He remained in Hawaii until July 1834, planning to return to England shortly, but he died on July 12. By that time his eyesight was failing, and he had already lost the sight of one eye. While walking along a well-marked path, he fell into a bull trap—a pit dug to catch a bull and covered over with vegetation to conceal it. The trap had already caught a bull. Douglas appeared to have fallen on top of the animal, and the bull killed him.

REGINALD FARRER AND ALPINE PLANTS

Rock gardens and the alpine plants that grow in them have been popular since the early years of the 20th century. They became fashionable following the publication in 1907 of *My Rock Garden*, a book that remained in print until about 1950. It was followed a year later by *Alpines and Bog Plants.* Both books were written by Reginald Farrer (1880–1920), a plant collector who made his first rock garden in an abandoned quarry at the age of 14 and helped make another at St. John's College, University of Oxford, while he was a student there.

Rock garden plants are known as alpines, which suggests that they originate in the European Alps. Many do, but others are Asian, and it was in Japan that Farrer acquired his ideas about garden design. In 1902 he spent eight months living in Tokyo, where he rented a house with a Japanese rock garden. He used his home as a base for brief excursions to Korea and China. This was his first botanical expedition, and his first book described it. Published in 1904, that book was called *The Garden of Asia.* Farrer later opened the Craven Nursery in his hometown of Clapham, Yorkshire, where he specialized in Asian mountain plants. The nursery closed during the economic troubles of the 1920s. Farrer believed that a rock garden should look as informal as possible and that its plants should appear to be growing in their natural habitat.

Reginald John Farrer was born in 1880 (the precise date is not recorded) at Ingleborough Hall, in the village of Clapham, North Yorkshire. He was born with a speech defect caused by a cleft palate and had to undergo many surgical procedures, as a consequence of which he was unable to attend school. Instead, he was educated at home until he entered St. John's College, University of Oxford, at the age of 17. He graduated in 1902.

On his return from Japan, Farrer's ambition was to be a novelist and poet, but his attempts were unsuccessful, and he turned instead to gardening. *My Rock Garden* and *Alpines and Bog Plants* both proved popular, and he followed them with *In a Yorkshire Garden* (1909) and *Among the Hills* (1910). He wrote *The English Rock Garden* in 1913, but its publication was delayed until 1919, after the end of World War I. When not writing, Farrer enjoyed walking and climbing in France, Switzerland, and Italy with friends who shared his enthusiasm for gardens. In 1907 he visited Sri Lanka, and at about that time he became a Buddhist. Farrer described his plant-hunting excursions in the Italian Dolomites in *The Dolomites: King Laurin's Garden,* published in 1913.

The following year Farrer embarked on a more ambitious expedition accompanied by his friend William Purdom (1880–1921), a

British plant explorer and botanist who had been trained at Kew Gardens. They spent two years exploring Tibet and Kansu Province of northwestern China, collecting hardy plants that were later introduced to British gardens. As always, Farrer's aim was to select plants that would grow in Britain with no need for costly heated greenhouses, so ordinary gardeners of modest means could enjoy them. Farrer told the story of this expedition in his two-volume work *On the Eaves of the World,* published in 1917.

As well as being a skilled field botanist and gardener, Farrer was also a talented painter. The Fine Art Society exhibited the watercolors he painted of landscapes in Tibet and Kansu in 1918, and he sent his paintings of plants to Sir Isaac Bayley Balfour (1853–1922), regius keeper at the Royal Botanic Garden Edinburgh, together with plant specimens and seeds. Balfour had a special interest in Sino-Himalayan plants.

Farrer's final expedition set off in 1919 to the mountains of Myanmar. This was less successful than his explorations in China and Tibet, because plants adapted to the tropical climate were mainly unsuitable for British conditions. Farrer died there, probably from diphtheria, on October 17, 1920.

There is still a fine display of Himalayan plants growing wild around Ingleborough Hall. These were sown by Farrer, sometimes by rather unorthodox methods. On one occasion he loaded a shotgun with seeds and fired them into a gorge and cliff near his home.

GEORGE FORREST, COLLECTING IN YUNNAN

George Forrest (1873–1932) was one of the most successful plant hunters of the early 20th century. He undertook seven major expeditions to western China and returned to the Royal Botanic Garden Edinburgh more than 30,000 specimens of 10,000 plant species, of which 1,200 were new to science. Forrest was also one of the most adventurous plant hunters and has been called "Scotland's Indiana Jones of the plant world." Many of the anemones, buddleias, asters, berberis, cotoneasters, and alliums now growing in private European and American gardens are descended from specimens Forrest collected, not to mention six species of rhododendron and more than 50 species of primula. Forrest was also a keen photographer. The illustration on page 99, from the Royal Botanic Garden Edinburgh, shows

George Forrest (1873–1932) was a Scottish plant hunter who undertook seven major expeditions to China, returning hundreds of species to Britain. *(Royal Botanic Garden Edinburgh)*

Forrest on one of his expeditions accompanied by his dog. The map on the next page shows the part of China where he worked.

Forrest was born on March 13, 1873, in Falkirk, Scotland. He was educated at Kilmarnock Academy and after leaving school was apprenticed to a druggist who used herbal treatments. As part of his training, Forrest learned medical botany, the uses of therapeutic plants, and the preparation of herbarium specimens. In 1891 he inherited enough money to allow him to seek a more exciting life, and he sailed to Australia to join the gold rush. He spent 10 years there, panning for gold and learning to survive in harsh conditions, before returning to Scotland in 1902, breaking his journey in South Africa.

Back in Edinburgh, Forrest found a job as a clerk in the herbarium of the Royal Botanic Garden, where Sir Isaac Balfour (see "Reginald Farrer and Alpine Plants" on pages 96–98) was impressed by his abilities. Balfour recommended Forrest for a plant-hunting expedition in China that was being planned, and he was accepted.

Forrest arrived at the town of Dali in Yunnan Province in August 1904 and spent some time getting to know the people and their customs and attempting to learn their language. In the summer of 1905, he set off with a team of 17 collectors for the northwestern corner of Yunnan, in the mountains close to the Tibetan border. They entered the rhododendron forests and collected large numbers of specimens and seeds. When they emerged from the forest, the political situation had deteriorated alarmingly. Warrior monks were torturing and killing foreigners and local people who had had any contact with foreigners. Forrest escaped, hiding by day and traveling by night, while parties of monks scoured the countryside in search of those they saw as offenders. Local people helped him hide, then helped him escape disguised in Tibetan dress. His ordeal did not prevent Forrest collecting. After a very brief time to recover,

Yunnan Province, in southwestern China, where George Forrest collected plants

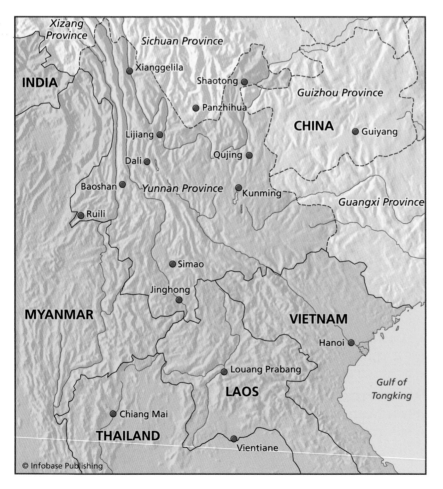

he continued his exploration in a different area, the valley of the Salween River (Nu Jiang), this time in the company of his friend George Litton, who was on the staff of the British consulate. They braved poisonous plants, clouds of biting insects, steep cliffs, and flimsy bridges spanning deep gorges. Not long after their return to Tengyueh in early January 1906, Litton died from malaria.

In March, Forrest set off again with a fresh team of collectors he had trained, this time to Lijiang in northern Yunnan. Now Forrest caught malaria and was forced to return to Dali, but his collectors continued working. In 1906 Forrest returned to Scotland with a huge collection of seeds, roots, tubers, plants, and herbarium specimens. The map of Yunnan opposite shows the region where Forrest worked.

Forrest returned to Yunnan six more times. His final expedition departed from Britain in 1930. Its purpose, he said, was to find all the plants that he had missed, and it did prove extremely productive. In January 1932 Forrest suffered a massive heart attack and died near the town of Tengyueh. He was buried on January 7 close to the grave of George Litton in the foreign cemetery outside the city of Tengyueh (now called Tengchong), to the west of Baoshan close to the Myanmar border.

ROBERT FORTUNE, COLLECTING IN NORTHERN CHINA

The Chusan palm (*Trachycarpus fortunei*) is one of the very few palm trees that can survive in the cool, wet Scottish climate. For that reason, it is a fairly common sight in Scotland, especially in coastal resorts attempting to offer a taste of the exotic to city vacationers. The species was introduced to Scotland by Robert Fortune (1813–80), who found it in 1844 growing on the Zhoushan (then transliterated as Chusan) archipelago, a group of 1,390 islands outside Hangzhou Bay just south of Shanghai. Fortune sent several of the plants to William Hooker at Kew (see "Sir William Hooker, the First Official Director" on pages 72–74) with a request that Hooker send one to Prince Albert's garden at Osborne House on the Isle of Wight. Legend has it that the tree arrived safely and that Queen Victoria planted it herself on her 32nd birthday, on May 24, 1851.

Other interesting and attractive plants grew on the Zhoushan Islands, including many azaleas, and Fortune spent the summer of

1844 there. The Chinese government had ceded Hong Kong to Britain in 1842 under the terms of the Treaty of Nanking, and Fortune had arrived there in July 1843 on a mission to collect plants for the Royal Horticultural Society (RHS) in London. The RHS had supplied their official collector in China with a list of requests. Fortune was to look for peonies with blue flowers, mandarin oranges, roses, azaleas, and tea plants, and he was to investigate the peaches that grew in the emperor's private garden, which were said to weigh two pounds (1 kg) each. The ornamental plants he found had to be sufficiently hardy to thrive in Britain. Fortune would collect some plants in the wild, but he would also aim to purchase plants already being cultivated in China—or acquire them in any other way he could. The RHS provided him with trowels and a spade, a cosh for self-defense, a double-barreled shotgun, and a Mandarin–English dictionary.

The English were not popular. The Treaty of Nanking was agreed at the end of the first of the opium wars in which the British forced the Chinese to import British opium. The treaty required the Chinese to pay heavy compensation for the opium their authorities had confiscated and destroyed to protect their population, to open several of their ports to British trading vessels, and, of course, to hand the important seaport of Hong Kong over to the British. It was highly unlikely, therefore, that Chinese growers would be willing to sell plants to Fortune, but he was resourceful. He learned to speak Mandarin well enough to pass as a person from a distant province with a curious accent. He wore Chinese dress and shaved his head to leave only a pigtail. His Chinese name was Sing Wah.

Fortune had many adventures. He survived storms and typhoons, was attacked by angry crowds who were not fooled by his disguise, and threatened by robbers and Yangtze River pirates. On one occasion he drove off a pirate attack by firing his shotgun at the two pirate junks. He shipped several consignments of plants back to Britain using the newly invented Wardian cases (see the following sidebar). The illustration that follows shows one of these cases, in which growing plants could survive for many months, allowing them to be shipped across the world. Fortune arrived back in London in May 1846 and became curator at the Chelsea Physic Garden. While there he wrote an account of his travels, *Three Years' Wanderings in the Northern Provinces of China*. This book was published in two volumes in 1847 and sold well.

THE WARDIAN CASE

Until the middle of the 19th century, plant hunters traveling in distant lands were able to send samples home only by preserving them first. Insects had to be killed and pinned to cards, and plants pressed and dried to make herbarium specimens. It was possible to send plant seeds on journeys lasting many weeks or even months, but there was no guarantee that they would germinate on arrival. By the latter half of the century the situation had changed. Collectors were able to send living specimens across the world, knowing they would reach their destinations in perfect condition.

The device that made this possible was the *terrarium*—a tightly sealed container for living plants and small land-dwelling animals. The terrarium was invented by Nathaniel Bagshaw Ward (1791–1868) and was more often called a Wardian case. Ward was a physician and possibly also a surgeon working in the East End of London, and he devoted his spare time to botany and entomology. He was particularly fond of ferns, but found these almost impossible to grow in his small garden owing to the amount of smoke and soot polluting the city air. Ward made his discovery when he tried to save a moth pupa by creating a natural environment for it inside a sealed jar. He did not record the fate of the moth, but after it had been in the jar for some time Ward noticed blades of grass and a fern growing in the soil at the bottom. He left them there, with the jar sealed, to find out how long they would survive, having realized that the plants were able to live because he had isolated them from the outside—and polluted—air. Ward wrote to Sir William Hooker at Kew Gardens describing what he had done, and he commissioned a carpenter to construct a container to his specification. This would comprise a wooden frame supporting glass windows on all sides and on top. He wished the frame to fit together as tightly as possible, and Ward stipulated that the hardest wood should be used, to prevent it decaying from the condensation and tropical warmth.

Two of the cases were delivered, and in July 1833 Ward filled them with local ferns and grasses and despatched them to Sydney, Australia. Six months later the cases arrived, and the plants inside them were thriving. Ward's Australian collaborator emptied the cases, cleaned them thoroughly, filled them with Australian plants, and in February 1835 they sailed for London. This time the voyage took eight months, and severe storms subjected the cases to a severe battering. Nevertheless, the plants arrived alive and growing well. Ward described his success in a small pamphlet, and in 1842 he published a book *On the Growth of Plants in Closely Glazed Cases*. His book attracted wide interest and his cases became very popular—and in the following years increasingly ornate. In 1854 Ward lectured on them to a meeting of the Royal Horticultural Society held at the Chelsea Physic Garden.

From that time, living plants could be sent around the world safely. Wardian cases were used to transport rubber trees from Brazil to Kew and from Kew to southern Asia, tea plants from China to India, and many ornamental plants from the Tropics to Europe. Ward became Master of the Society of Apothecaries, a fellow of the Linnean Society, and a fellow of the Royal Society.

Fortune's second expedition to China was made in 1848 on behalf of the British East India Company. The company wished to establish tea plantations in India, and for that purpose they asked Fortune to obtain shrubs of the highly prized silver-tipped tea, as well as instruc-

The Wardian case, invented in about 1829 by Dr. Nathaniel Bagshaw Ward (1791–1868), made it possible to transport living plant specimens around the world and have them arrive in a healthy condition.

© Infobase Publishing

tions on their cultivation. At that time China was the only source of the best quality tea, and the Chinese were determined to keep their monopoly—the penalty for smuggling tea was death by beheading. The shrubs grew in the north of Fujian Province. Fortune disguised himself as an official from the imperial palace, and with the help of local workers over the course of three years he was able to collect more than 20,000 plants and send them to northern India in Wardian cases. Tea seeds cannot be stored, so living tea plants had to be shipped. The shrubs were planted in the hills around Darjiling (then called Darjeeling). A number of Chinese workers also made their way to Darjiling to instruct local people in the arts of tea growing. Tea lovers consider Darjeeling one of the finest of all teas; it is sometimes called the champagne of teas. An Indian tea industry prospered, and the Chinese monopoly was broken.

Fortune returned to China for two further expeditions, for the British East India Company in 1853–56 and for the U.S. Patent Office in 1858–59 to collect seeds of plants that might be cultivated in the United States. From 1860 to 1862, Fortune was acquiring plants in Japan on his own behalf. On that expedition he did not search for wild plants, but bought cultivated varieties, including chrysanthemums, from Japanese nurseries. He also visited Indonesia, Taiwan, and the Philippines. In all, Fortune was responsible for introducing more than 120 species of garden plants to Britain. He described his expeditions in *A Journey to the Tea Countries of China* (1852), *Two Visits to the Tea Countries of China and the British Tea Plantations in the Himalaya* (1853), *A Residence Among the Chinese* (1857), and *Yedo and Peking: A Narrative of a Journey to the Capitals of Japan and China* (1863).

Robert Fortune was born on September 16, 1812, at Edrum, near Duns, Berwickshire, in the Scottish borders. He was educated locally, and after leaving school he served an apprenticeship to a gardener in the nearby village of Kelloe. His apprenticeship completed, he went to work at the Royal Botanic Garden Edinburgh. In 1842 he obtained the post of deputy superintendent of the hothouse department at the Royal Horticultural Society at Chiswick, London. He retired after his return from Japan in 1862, returning to Scotland and becoming a farmer in East Lothian. His books brought him an income that allowed him to live comfortably. He died on April 13, 1880.

ERNEST WILSON, COLLECTING IN CHINA AND JAPAN

In April 1902 Ernest "Chinese" Wilson (1876–1930) arrived back in England from his first plant-collecting expedition. He brought with him 35 Wardian cases of plants (see the sidebar "The Wardian Case" on page 103), seeds of 305 plant species, and herbarium specimens of 906 species. Between 1899 and 1919, Wilson undertook five expeditions to China and three to other parts of the world, from which he sent back to Britain and the United States more than 1,000 garden plants—perhaps as many as 2,000—and about 16,000 herbarium specimens.

Wilson's 1902 haul included *Actinidia deliciosa.* This plant is known in China as mihoutau and has been cultivated for many centuries for its fruit, which is used in many traditional recipes and in Chinese medicine. Robert Fortune (see "Robert Fortune, Collecting in Northern China" on pages 101–105) had sent specimens of it to London, but *Actinidia* plants are *dioecious*—they have separate male and female plants—and Fortune collected only female plants, so those grown in England were incapable of fruiting. Wilson found mihoutau growing at Yichang, in Jiangxi Province, where he had established his base and where there was a small community of foreigners to whom Wilson warmly recommended it. Isabel Fraser (1863–1942), a New Zealand schoolteacher and evangelist at the Church of Scotland mission in Yichang, shared Wilson's enthusiasm. When she returned home in 1904, she took some seeds with her and gave them to a family of horticulturists. The result grew into a global industry based on improved varieties. There were problems with the name. For a time the fruits were known as Chinese gooseberries, but when they were marketed in the United States that name was deemed inappropriate so the New Zealanders exporting them thought to give them a distinctively New Zealand–sounding name. They called them kiwifruit.

Ernest Henry Wilson was born in the small town of Chipping Campden, Gloucestershire, on February 15, 1876. He attended a local school, and after leaving he found work as an assistant gardener at Hewitts Nursery in Warwickshire. When he was 16, Wilson moved to the Birmingham Botanical Gardens, and while working there he studied botany in the evenings at Birmingham Technical School, where he won the Queen's Prize for botany. In 1897, aged 21, Wilson commenced work at the Royal Botanic Gardens, Kew. While there, he won the Hooker Prize for an essay on conifers and began teaching botany at the Royal College of Science (which later became part of Imperial College of Science and Technology of the University of London).

In the late 19th century James Veitch and Sons ran the largest chain of plant nurseries in Europe, and they approached Wilson with a proposal to send him to China in search of just one plant—the dove tree (*Davidia involucrata*), also known as the handkerchief tree because when it is in flower the large, pure white *bracts* hang down, looking like handkerchiefs. Sir Harry Veitch (1840–1924), head of the company, employed a team of plant hunters, and he advised Wilson to "Stick to the one thing you're after, and don't spend time and money wandering about. Probably every worthwhile plant in China has now been introduced to Europe." Wilson spent six months at the Veitch nursery before setting off for China.

Wilson traveled west and spent five days at Harvard University's Arnold Arboretum in Boston, Massachusetts. He had a letter of introduction to the director, the distinguished American botanist Charles Sprague Sargent (1841–1927), who taught him how to prepare plant specimens for long-distance transport. He also visited Sargent's own gardens on his estate at Brookline. Wilson crossed the United States by rail and sailed from San Francisco to Hong Kong, where he arrived on June 3, 1899. He found the dove tree and spent two years collecting plants, mainly in the remote mountain valleys of Hupeh Province.

On his return to England in 1902 Wilson married Helen Ganderton, but it was not long before Veitch sent him back to China, this time in quest of the Chinese poppy (*Meconopsis integrifolia*). Wilson spent his time in western Szechuan Province, where one of the plants he found in 1903 was the regal lily (*Lilium regale*).

In 1907 Sargent asked Wilson to go back to China to find trees and shrubs suitable for American gardens. He returned to Szechuan again in 1908 and 1910, also on behalf of the Arnold Arboretum. On each visit he sent lily bulbs back to Sargent, but most of them rotted on the journey, and it was not until his third attempt that viable bulbs reached their destination and the lilies were introduced to North America. During the 1910 expedition, a rock fall crushed Wilson's leg. He set the fracture using his camera tripod as a splint, but for the rest of his life he walked with what he called his "lily limp."

Wilson collected cherry trees in Japan from 1911 to 1916, working for the Arnold Arboretum, and he returned with 63 varieties. In 1917 and 1918 he was hunting plants in Korea and Taiwan (then called Formosa). On his return to the United States in 1919, he was made associate director of the Arnold Arboretum. In 1921 he embarked on a two-year expedition to Australia, New Zealand, India, South America, Central America, and East Africa. In 1927 he was appointed keeper of the Arnold Arboretum. He received many honors. The Royal Horticultural Society awarded him the Victoria Medal of Honor and the Veitch Memorial Medal. The Massachusetts Horticultural Society presented him with the George Robert White Memorial Medal. He was made a fellow of the American Academy of Arts and Sciences. Harvard University awarded him an honorary M.A. degree, and Trinity College, Connecticut, gave him a D.Sc. degree.

On October 15, 1930, Wilson and his wife were driving near Worcester, Massachusetts, when their car skidded on a slippery road and fell 40 feet (12 m) down an embankment. They were both killed.

Geography of Plants

During the 18th century, as specimens of exotic plants arrived in Europe from every corner of the world, it became evident that particular types of plant were being found in particular places. Similar types of vegetation occurred in each of the continents apart from Antarctica, but each continent had its own distinctive species making up that vegetation. Tropical forests grow in Central and South America, Africa, Asia, and Oceania, for instance, and the forests appear to be very similar, but the trees, shrubs, and other plants growing in them are not the same. Certain euphorbias growing in arid regions of Africa closely resemble cacti, but they are not related and except for one species (*Rhipsalis baccifera* found in Africa, Madagascar, and Sri Lanka) cacti occur naturally only in America. Naturalists became interested in mapping the geographic distribution of plant species and the new science of *biogeography* was born.

This chapter outlines the development of plant geography. It begins with the story of the Prussian aristocrat who became one of the world's most famous explorers and continues with a brief account of how the map of the world was divided into botanical regions. The chapter ends with the story of the Swedish botanist who was one of the most important plant geographers of the early 20th century.

ALEXANDER VON HUMBOLDT AND THE PLANTS OF SOUTH AMERICA

Alexander von Humboldt's exploration of South America, accompanied by the French botanist Aimé Bonpland (1773–1858), did much to establish biogeography as a distinct scientific discipline, but Humboldt was much more than a geographer. In his most famous work, *Kosmos*, published in five volumes between 1845 and 1862, he wrote that "I always wanted . . . to understand nature as a whole . . . the separate branches of natural knowledge have a real and intimate connection." He was an Earth scientist who founded a school of mining and invented a safety lamp for miners. He studied volcanoes, measured the Earth's magnetic field, and discovered the Peru Current off the Pacific coast of South America—it is sometimes called the Humboldt Current. He was a professional diplomat, a fervent opponent of slavery, and a supporter of self-determination for the citizens of European colonies. The illustration at left is from a portrait painted in 1806, two years after his return from South America, when he was 37 years old.

In June 1802 Humboldt and Bonpland, accompanied by Carlos Montúfar (1780–1816), climbed Mount Chimborazo in Ecuador. This volcano, 20,569 feet (6,269 m) high, was thought at the time to be the world's highest mountain, and the group reached a height of 18,893 feet (5,762 m) before mountain sickness defeated them. Humboldt recognized that it was lack of oxygen that caused the illness, but there was nothing they could do but descend. Nevertheless, when news of their climb reached Europe it made them into celebrities.

They did not simply climb the mountain, however. As they went they recorded the plants growing on the mountainside and their locations. Later Humboldt drew a cross section of the mountain, with the names of the plants at their correct altitudes written onto the drawing. He added two tables, one on each side of the main drawing, in which he recorded the temperature, barometric pressure, and other climatic details at each elevation. That drawing clearly demonstrated his belief

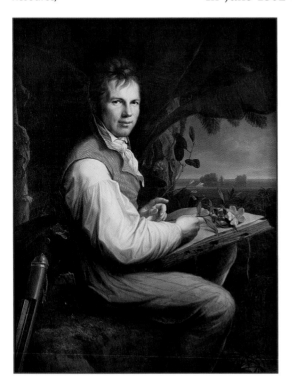

Alexander von Humboldt (1769–1859) was a Prussian naturalist and geographer who explored Central and South America. This portrait, painted in 1806, two years after his return, shows him studying a plant specimen. *(Bildarchiv Preussischer Kulturbesitz/Art Resource)*

in the interconnectedness of natural phenomena. During their exploration of what was then Spanish America, Humboldt and Bonpland collected 5,800 species of plants, of which 3,600 were unknown to science until that time. They—but mainly Bonpland—identified and classified most of them before leaving the United States.

Humboldt's full name was Friedrich Wilhelm Heinrich Alexander, baron (Freiherr) von Humboldt, and he was born in Berlin, then the capital of Prussia, on September 14, 1769. He was taught by tutors before entering the University of Frankfurt-an-der-Oder in 1789 to study economics, followed by a year at the University of Göttingen, where he developed the interest in natural history he had shown since he was a young child. In 1791 Humboldt enrolled at the Freiberg School of Mining (now the Technische Universität Bergakademie Freiberg). He left in 1792 without taking a degree and was appointed inspector of mines.

Following the death of his mother in 1796, Humboldt inherited a considerable sum of money. In 1797 he resigned from the Prussian department of mines and went to Paris, where he met and became friendly with the French botanist Aimé Bonpland. The two friends traveled to Madrid, where the Spanish prime minister obtained official permission for them to visit Spain's American colonies. They sailed from Marseille and arrived at Cumaná, New Andalusia (now Venezuela), on July 16, 1799. By the end of their exploration in 1804, Humboldt and Bonpland had traveled more than 6,000 miles (9,650 km) on foot, horseback, and by canoe. The following map shows the route they followed.

Humboldt lived in Paris from 1804 until 1827, writing his account of the travels. His *Essay on the Geography of Plants* appeared in 1805 and became volume 5 of his 23-volume *Personal Narrative of Travels to the Equinoctial Regions of the New Continent,* coauthored with Bonpland, the final volume of which was published in 1824. He spent his final years working on *Kosmos,* his greatest masterpiece, which developed from a series of lectures he delivered in Berlin in 1827–28. In it he aimed to bring together all of the sciences in a comprehensive portrait of the Earth. The first two volumes were published between 1845 and 1847, volumes 3 and 4 between 1850 and 1858, and as much of volume 5 as he was able to complete appeared in 1862 after his death. Humboldt suffered a minor stroke on February 24, 1857, from

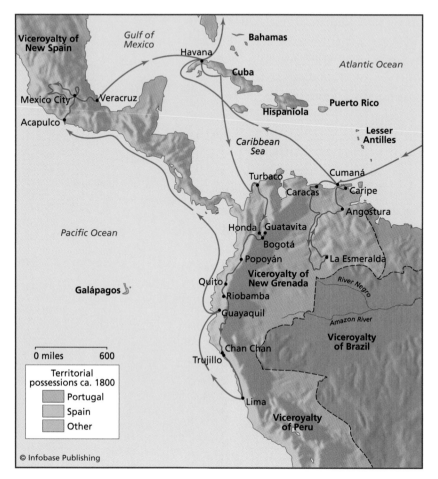

The route Alexander von Humboldt and Aimé Bonpland followed through Central and South America between 1799 and 1804

which he recovered, but his strength weakened during the winter of 1858–59, and he died peacefully on May 6, 1859. He was given a state funeral.

Aimé-Jacques-Alexandre Goujaud was born in La Rochelle, France, on August 28, 1773. He later changed his name to Bonpland. He studied medicine in Paris from 1791 to 1794, served as a surgeon in the French army until 1795, and then resumed his medical studies. On his return to France in 1804, Bonpland was awarded an official pension and made director of the empress Josephine's private botanical garden at Malmaison, her country house about seven miles (11 km) from Paris, where he spent much of his time arranging his botanical collections and writing. In 1808 Josephine appointed him her official botanist. In 1813 Bonpland published *Description des Plantes Rares Cultivée à Malmaison et à Navarre* (Description of the rare plants cultivated at Malmaison and Navarre), a book describing the contents of Josephine's garden. Josephine died in 1814, and in 1816 Bonpland returned to South America to take up a post as professor of natural sciences in Buenos Aires, Argentina, where he also practiced medicine. In 1817 he was made a corresponding member of the French Académie des Sciences. In 1820 he left Buenos Aires to establish and run a plantation near the Paraná River, but he offended the dictator of Paraguay, who had him detained until 1829. After his release, Bonpland started other plantations in Brazil and Uruguay, which he managed from 1831 until his death. In 1854 Bonpland was decorated by King Friedrich Wilhelm III of Prussia and in 1856 he received an honorary degree from the University of Greifswald, in Germany. Bonpland died on May 11, 1858, at Restauración, Argentina.

KARL LUDWIG VON WILLDENOW AND THE START OF SCIENTIFIC PLANT GEOGRAPHY

In 1788 when he was 19 years old and about to enroll at the University of Göttingen, Alexander von Humboldt met Karl von Willdenow (1765–1812), a 23-year-old medical student. Humboldt already had a keen interest in natural history and his meeting with Willdenow strongly encouraged it. Willdenow was learning medical botany, and his detailed studies of plants had led him to consider how they were distributed geographically.

Willdenow was not the first naturalist to observe patterns in the way plants were distributed. Johann Reinhold Forster (1729–98) had made a similar observation earlier. Forster, with his son Georg (1754–94) as his assistant, had been the official naturalist on James Cook's (1728–79) second voyage in search of a southern continent, from 1772 to 1775. Johann Forster was a difficult man. After their return to England, he quarreled with both Cook and the Admiralty over who should write the official account of the voyage. The Admiralty finally backed Cook, leaving Forster to publish his own account at his own expense. In fact, he wrote two. The first, written in collaboration with Georg and published in 1777, was entitled *A Voyage round the World in His Britannic Majesty's Sloop* Resolution, *Commanded by Capt. James Cook, during the Years 1772, 3, 4, and 5.* The following year he published *Observations Made During a Voyage Round the World,* of which he was the sole author. In that book, Forster noted that within each region of the world the expedition visited, the plants and animals formed distinct units, each occupying a particular environment defined principally by temperature. As they sailed from the frigid Antarctic to the Tropics, these biological units formed a clearly defined sequence, and he noted that the tropical plants and animals were more abundant and more spectacular than those found anywhere else. Forster also recorded that the species living on islands were similar, though not identical, to those found on the nearest continent, but that Asian and American species mingled on some Pacific islands. It was the beginning of biogeography, but Forster unjustifiably extended his observations of plants and animals to human populations.

In later years, when he had become a professional botanist, Willdenow concentrated on plant distribution and avoided Forster's vague anthropological speculations. He recognized that climatic, geological, and biological factors interacted to produce regional differences in plant communities. These differences were due to plants adapting to the conditions in which they lived, but Willdenow saw that this was not a complete explanation. In many instances the plants living under similar conditions in different regions were different species and not the same species that had adapted in different ways. Willdenow saw that the climate was the most important factor determining the type and number of plant species in a particular region. He also noted that the composition of plant communities changed over time, and that occasionally plants could cross geographic barriers such as mountains and oceans, thereby expanding their range from one region to another. He observed that new species could appear and that others could become extinct. Willdenow published his most important book in 1792. It was entitled *Grundriss der Kraüterkunde zu Vorlesungen* (translated into English and published in 1805 as *Principles of Botany*), and it was the book in which Willdenow laid the foundation of plant geography.

Karl Ludwig von Willdenow was born in Berlin on August 22, 1765, the son of an apothecary. He studied pharmacy at Wieglieb College in Bad Langensalza, Thuringia, graduating in 1785 and then enrolling at the University of Halle, where he studied medicine and botany. He graduated in medicine in 1789, and the following year he took over his father's apothecary business. Willdenow worked as an apothecary until 1798, but combined this with his continuing study of plants and their distribution. In 1794 he became a member of the Berlin Academy of Sciences. In 1798 he was appointed professor of natural history at the Berlin Medical-Surgical College, and in 1810 he was made professor of botany at the University of Berlin. In 1801, the year he became director of the Berlin Botanical Garden, he was appointed principal botanist to the Berlin Academy of Sciences.

Willdenow had the botanical garden replanted, grouping the plants according to the part of the world and the type of habitat where they were found. Plant collectors, including Humboldt, sent him specimens from all over the world, and he assembled a large herbarium, which the botanical garden purchased in 1818 after his death. In 1807 Humboldt obtained funding for Willdenow to expand the garden. Willdenow remained director of the botanical garden until his death in Berlin on July 10, 1812.

FRANZ MEYEN AND VEGETATION REGIONS

Humboldt was generous in his support for young researchers and helped many at the start of their careers, but if he can be said to have had a favorite, that person was Franz Meyen (1804–40). When Humboldt first met him, Meyen was working as a physician. Humboldt secured for him the post of professor of botany at the University of Berlin and remained close to him. For his part, Meyen followed in Humboldt's footsteps, exploring South America from 1830 to 1832 and making scientific observations during climbs in the Andes.

His travels led Meyen to develop his own ideas on plant geography, which he published in 1834 as *Grundriß der Pflanzengeographie* (Outline of plant geography). In this work Meyen showed the influence of climate, soil, and other environmental factors on the type of vegetation, and he used *isolines*—lines joining places with similar values for a specified variable, first used by Humboldt—to delineate the boundaries of regions with a distinctive type of vegetation. He also discussed in the book the origin and spread of cultivated plants and their uses. Humboldt did much to promote Meyen's book.

Meyen was a brilliant botanist, with an interest in every aspect of the subject. In the same year as he published *Grundriß*, Meyen also published a book entitled *Über die neuesten Fortschritte der Anatomie und Physiologie der Gewächse* (On the latest research into the anatomy and physiology of plants). His book *Phytotomie*, published in 1830, was one of the first works on the microscopic anatomy of plants. He developed a cellular theory of plant structure, which he outlined in a book entitled *Untersuchungen über den Inhalt der Pflanzen-Zellen* (Inquiries into the contents of plant cells), published in 1828.

Franz Julius Ferdinand Meyen was born at Tilsit, Prussia (now Sovetsk, Russia), on June 28, 1804, the son of a bookkeeper who worked in a small store. Meyen left high school early and began studying pharmacy at Memel, East Prussia (now Klaipeda, Lithuania), but in 1821 he moved to the University of Berlin and changed to medicine, in which he qualified in 1826. He practiced for four years as an army doctor and as a physician in Berlin, Cologne, and Bonn, before setting off on his exploration of South America on board the *Prinzess Luise.* He visited Brazil, Peru, and Bolivia and then crossed the Pacific, visiting a number of islands including Oahu, Hawaii. He spent a short time in China and at the island of St. Helena. On his return to Germany he was appointed professor of botany at the University of Berlin, where he remained until his death in Berlin on September 2, 1840.

ALPHONSE DE CANDOLLE AND WHY PLANTS GROW WHERE THEY DO

Augustin de Candolle (see "Augustin de Candolle and Natural Classification" on pages 86–88) had traveled in Brazil, Indonesia, and northern China, and his observations convinced him of the importance of soil type in determining the way plants are distributed. In 1820 he wrote *Essai élémentaire de géographie botanique* (Elementary essay on botanical geography), in which he elaborated on an idea that Linnaeus (see "Carolus Linnaeus and the Binomial System" on pages 83–86) had proposed earlier.

Linnaeus—who did not believe that species evolve from earlier species—had maintained that all plants and animals originated from the same particular place. Many scientists then believed that at one time oceans had covered the entire Earth. Linnaeus proposed that plants and animals first appeared on a mountain that was high enough to protrude above the ocean surface, forming an island close to the equator. The mountain provided a variety of living conditions, varying with elevation, and the different species of plants and animals inhabited those parts of the mountainside that best suited their needs. As the waters receded, pairs of animals and plants migrated to the regions of the newly exposed dry land that replicated their original habitats and there they remained, forever unchanging. This meant, according to Linnaeus, that there were two aspects to the study of plant distribution. The first was the "study of stations." This meant the physical causes for a plant species growing in a particular place or the features that made a locality especially suitable for that plant. The second was the "study of habitations." This concerned geographical or geological factors that affected the overall range of a

plant species. These factors might no longer exist, but they had originally defined the range. Candolle concentrated on what those factors
might be and how they might influence distribution. He believed
it possible that certain species possessed such a large capacity for
dispersal that they could spread from a station to a habitation and
eventually achieve a global distribution.

Augustin also taught his son Alphonse (1806–93) plant geography, and Alphonse de Candolle continued his father's work. That
work was partly concerned with plant taxonomy, but Alphonse was
also keenly interested in plant geography and analyzed in detail the
environmental factors, especially temperature, influencing plant
distribution. He published his findings in 1855 in his two-volume
treatise entitled *Géographie botanique raisonnée* (Analytical botanical geography). In this work Candolle recognized 20 distinct botanical regions, as well as the distinct floras found on islands. Candolle
was a skilled statistician who used statistical techniques to measure
the global distribution of vegetation types. This led him to define the
types of plants dominating the vegetation in particular types of climatic environment. The climatic environments, determined mainly
by temperature, divided the world into latitudinal belts of arctic,
temperate, and tropical vegetation. In 1874, in one of the volumes
of *Prodromus systematis regni vegetabilis* (Introduction to a natural
classification of the vegetable kingdom), he gave the following names
to the resulting vegetation types:

Hekistotherms—typical of arctic tundra
Microtherms—typical of cool temperature deciduous forest, cool
    temperate coniferous forest, and *boreal forest*—the coniferous
    forest of subarctic climates that forms a belt across northern
    North America and Eurasia
Mesotherms—typical of warm temperate deciduous forests, warm
    temperate coniferous forests, and Mediterranean climates
Xerophiles—typical of deserts and grasslands
Megatherms—typical of tropical rain forest and tropical savanna

It was from Alphonse de Candolle that Charles Darwin (see
"Charles Darwin and Evolution by Means of Natural Selection" on
pages 151–154) learned about plant geography, and they conducted
a detailed correspondence on various aspects of plant distribution.

Candolle's classification of vegetation types also represented an approach to climate classification, and in 1884 the German meteorologist and climatologist Wladimir Peter Köppen (1846–1940) used Candolle's vegetation types in his first attempt at defining climatic zones. The final version of the Köppen climate classification was published in 1946, after Köppen's death, and it is still the most widely used classification.

Alphonse Louis Pierre Pyrame de Candolle was born in Paris on the night of October 27/28, 1806. In 1816 the family moved to Geneva. Candolle studied law at the University of Geneva, graduating in 1825. For most of his life, starting in 1824 when he was still a student, Candolle continued work commenced by his father. In 1829 Candolle received a degree in law and in 1831 he was appointed an honorary professor at the Academy of Geneva. From 1835 to 1850 Candolle was professor of botany at the University of Geneva and director of the university's botanical garden. He retired from his teaching commitments in 1850 to give more time to research. Candolle devised the first code of botanical nomenclature, which was adopted in 1867 at the International Botanical Congress that Candolle had organized in Paris. He published it in the same year with the title *Lois de la nomenclature botanique adoptées par le Congrès international de botanique tenu à Paris en août 1867* (Laws of botanical nomenclature adopted by the International Botanical Congress held in Paris in August 1867). The present International Code of Botanical Nomenclature is its direct descendent.

Candolle was elected to the French Academy of Sciences in 1851, to the Royal Society of London as a foreign member in 1869, and to the National Academy of Sciences of the United States in 1883. He died in Geneva on April 4, 1893.

EDWARD FORBES AND THE SIGNIFICANCE OF ICE AGES

The River Tees rises in the northern Pennines—a chain of hills that runs in a north-south direction down the center of northern England—flowing about 70 miles (113 km) to the North Sea. In the upper part of the valley, called Upper Teesdale, there is an unusual community of plants. The plants are not rare, but ordinarily they occur only in arctic or alpine environments very much higher in

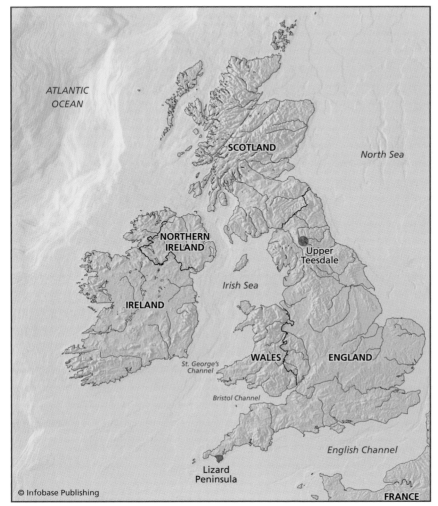

In Upper Teesdale and on the Lizard Peninsula there are plant communities that survive from a time when the climate was very different from that of today.

latitude or elevation than Upper Teesdale. How do they come to be where they are?

In the far southwest of England a peninsula projects into the ocean. Its most westerly point is called Land's End and its most southerly point—and the most southerly point of mainland Britain—is Lizard Point at the tip of the Lizard Peninsula. On the Lizard Peninsula there is another unusual community of plants. In this case the nearest places with a similar group of plants are in southern and western Ireland, and the plants occur more extensively in Portugal. The map above shows the locations of Upper Teesdale and the Lizard

Peninsula. Why are isolated patches of Iberian plants growing in Ireland and southern Britain?

During the first half of the 19th century, Sir Charles Lyell (1797–1875) was a dominant figure in British geology. One of the great geological debates at the time concerned the history of the Earth, on which there were two views. Catastrophists believed the present Earth had been shaped by a series of extremely violent events, while uniformists believed that the Earth had developed through the slow and continual operation of processes that were still taking place. Lyell was a uniformist. He argued that very gradually, over an unimaginably long period, the Earth's surface changed, and as it did so some areas of habitat were destroyed, others greatly modified, and new habitats formed. This, he maintained, allowed species to migrate as widely as they could into new territories, only to become separated from their earlier territories by the appearance of barriers such as mountains and oceans. The barriers were temporary, however, and after millions of years they disappeared as the land was restructured. Lyell did not believe in the evolution of species. He held that each species was created once, in a single event and in a particular place. The distribution of species was therefore to be explained as the result of migrations from certain centers of creation.

Hewett Cottrell Watson (1804–81) was an English botanist with a keen interest in plant distribution. In 1832 he published *Outline of Distribution of British Plants,* the first of several studies he made of plant distribution in Britain. Watson later divided Great Britain into 112 and Ireland into 40 named areas that he called vice counties, publishing this scheme in 1852 in his book *Cybele Britannica.* This was the beginning of the compilation of a systematic census of British flora. Watson also observed the way that the composition of plant communities changed with altitude.

Watson described plant communities, but he did not speculate about how they came to be distributed. The Manx geologist Edward Forbes (1815–54) had no such inhibition, however, and combined Watson's vice counties with Lyell's work on Earth history to reach what he believed to be a plausible explanation of distribution. Forbes set out his new theory in 1846, in an essay entitled "On the connexion between the distribution of the existing Fauna and Flora of the British Isles, and the Geological Changes which have affected their Area, especially during the epoch of the Northern Drift," that appeared in the *Memoirs of the Geological Survey of Great Britain.* He returned to this theme at the 1854 annual meeting of the British Association for the Advancement of Science, held in Cambridge, where he delivered a paper "On the geographical distribution of local plants." Later he delivered a series of six lectures "On the geographical and geological distribution of organized beings" at the Royal Institution in London.

Forbes divided the British Isles into five botanical zones and proposed that the plants and animals inhabiting each of these zones had migrated to them from different regions of continental Europe. The plants of western and southwestern Ireland were related to those in northern Spain, those of southeastern Ireland and southwestern England to plants of the Channel Islands and adjacent France, the plants of southeastern England to those on the opposite side of the English Channel, those on mountains to plants in arctic Scandinavia, and the remainder to those in Germany. Forbes calculated that there had been three distinct periods of migration. An area of land had at one time been exposed in the eastern Atlantic, allowing Iberian plants to migrate northward; at various times the English Channel had disappeared, and land bridges had appeared and disappeared. The mountain flora had drifted southward, carried on icebergs at a time when temperatures were much lower than they were in the 19th century and a large part of northern Europe had been submerged beneath a shallow sea, with the mountains standing above the surface as islands. When the sea level fell and temperatures rose, the plants migrated to higher elevations.

Edward Forbes was born on February 12, 1815, in Douglas, Isle of Man. Natural history fascinated him from an early age, but his health was too delicate for him to attend school until 1828, when he entered Athole House Academy in Douglas as a day pupil. He moved to London in June 1831, intent on becoming an artist, but gave up the idea and returned to Douglas in October. In November he began to study medicine at the University of Edinburgh, but by 1836 he had abandoned any idea of a medical career and instead devoted himself to literature and science. Between 1832, when he examined the plants and animals of the Isle of Man, until 1842, Forbes made a number of field excursions and visits to museums in various parts of Europe. In 1842 he was appointed professor of botany at Kings College London and also as a curator for the Geological Society. In 1844 the Geological Survey appointed him as a paleontologist. He became president of the Geological Society and of the geological section of the British Association for the Advancement of Science in 1853. In 1854 he was made Regius Professor of Natural History at the University of Edinburgh. Forbes was elected a fellow of the Linnean Society in 1843 and of the Royal Society in 1845. He died after a short illness at Wardie, near Edinburgh, on November 18, 1854.

Forbes came very close to a satisfactory explanation for the distribution of British plants and also for the arctic-alpine plants of Upper Teesdale and the warm-temperate plants of Ireland and the Lizard Peninsula. *Paleoclimatologists*—scientists who study the ancient history of climate—have discovered that the Upper Teesdale plants are survivors from a period of much colder climate during the most recent ice age. The Irish and Lizard plants are probably survivors from the time when temperatures were rising as the last ice age came to an end. At that time sea levels were much lower than those of today, because of the amount of water held in the ice sheets that covered much of northern Europe and North America. Both the Irish Sea and the English Channel were dry land, and the plants most likely migrated northward from Portugal. It is also possible, but there is no conclusive evidence, that these plant communities have survived even longer, from the period preceding the last ice age.

AUGUST GRISEBACH AND FLORAL PROVINCES

In 1872 August Grisebach (1814–79), professor of botany at the University of Göttingen, Germany, published a two-volume work that greatly advanced the study of phytogeography. Its full title was *Die Vegetation der Erde nach ihrer klimatischen Anordnung: Ein Abriss der vergleichenden Geographie der Pflanzen* (The vegetation of the Earth according to its climatic distribution: A sketch of the comparative geography of plants). Grisebach was one of a number of geographers who had been inspired by the travels of Alexander von Humboldt (see "Alexander von Humboldt and the Plants of South America" on pages 110–113). In 1839 and 1840 Grisebach traveled through the Balkans in southeastern Europe and Turkey studying the geography and the plants. His account of these and other travels established his scientific reputation. He devised the term *Geobotanik* (geobotany) to describe his work.

Humboldt had drawn *isotherms*—lines joining points on the Earth's surface or at the same elevation with respect to the surface, where the temperature is the same—across maps of the world. These defined the boundaries of climatic—and therefore vegetational—zones. For Humboldt's followers, including Grisebach, this offered a new approach to the study of plant distribution. Rather than listing the plants of a region and then trying to figure how they came to be there, the Humboldtian alternative was to examine the ways in which plants were adapted to the climatic conditions in which they lived. This led Grisebach to the realization that communities occurring in a particular climate were similar, even though the species composing them were different. North American prairie, for instance, was very similar to European steppe and the South American pampas, but each type of grassland supported its own species of grasses and herbs. In the same way, humid tropical forests were similar whether they occurred in South America, Africa, or southern Asia, although each had its own suite of tree species. Grisebach called an assemblage of plants determined by climate a *formation*. This was a new way of looking at plants. They could now be studied as communities, opening a new branch of plant science that led later to the discipline of *phytosociology,* which is the classification of plant communities according to the characteristics and relationships of and among the plants within them.

Grisebach used this approach to divide the world into *floral provinces.* A floral province is a group of plants covering a large geographic area, all of which are adapted to the climate of that area. Grisebach shared the view of Humboldt, Lyell, Forbes, and others that plant communities had come into existence within the provinces where they were found, quite independently of events elsewhere.

August Heinrich Rudolf Grisebach was born in Hannover, Germany, on April 17, 1814. He studied medicine and botany from 1832 to 1836, first at the University of Göttingen, where his uncle, Georg Friedrich Wilhelm Meyer (1782–1856), was a professor of botany, and later at Berlin, where he received his doctorate in medicine in 1836. In 1833, while still a student, he traveled in the European Alps studying the vegetation. After qualifying, Grisebach moved to Göttingen as a Privatdozent—a qualified academic who teaches without being paid by the university but receives payment from his students. In 1841, following his return from his travels through the Balkans and Turkey,

Grisebach was appointed associate professor of botany within the medical faculty at Göttingen. He became a full professor in 1847, and in 1875 he was appointed director of the Göttingen botanical garden. Grisebach died at Göttingen on May 9, 1879.

CARL SKOTTSBERG AND THE PLANTS OF SOUTHERN SOUTH AMERICA

At Punta Arenas, in southern Chile, there is a botanical garden named in honor of one of the world's greatest botanical explorers. The Jardin Botanico "Carl Skottsberg" was founded in 1970, just a few years after the death of the Swedish botanist Carl Skottsberg (1880–1963). Early in his career, Skottsberg took part in a Swedish expedition to Antarctica, where he studied the vegetation on islands off the Antarctic coast. A few years later he led an expedition to study the distribution of plants in Patagonia, Tierra del Fuego, the Falkland Islands, South Georgia, and the islands of Juan Fernandez, which are much farther north, at latitude 33.25°S.

Skottsberg devoted decades to the study of the flora of the far south. He concluded that before its surface vanished beneath the ice, Antarctica supported plants and animals that migrated northward, crossing land bridges that once linked Antarctica with South America, Australia, and New Zealand. He also thought it possible that much earlier a land bridge had existed between Antarctica, the Kerguelen Islands, and South Africa. His investigation of the flora of Juan Fernández, off the Chilean coast, led him to propose that long ago there had been land extending northward from Antarctica parallel to the present South American coastline. Plants and animals moved northward along this strip of land, eventually reaching Juan Fernandez and the adjacent volcanic islands of Más a Fuera and Más a Tierra. This meant that the flora of these islands was a relic of climatic conditions that had existed in the remote past, like the floras of Upper Teesdale, Ireland, and the Lizard Peninsula. The modern view is that continental drift, driven by the movements of the plates making up the Earth's crust, transports landmasses together with their living inhabitants, but Skottsberg did not accept the theory of continental drift.

Carl Johan Fredrik Skottsberg was born on December 1, 1880, at Karlshamn, in southern Sweden. His father, Carl Adolf Skottsberg, was rector of the local boys' school. Skottsberg entered the University of Uppsala in 1898 and obtained his doctor's degree in 1907. He became a lecturer at the university the year he received his doctorate, and in 1909 he was made keeper of the herbarium at the Uppsala Botanical Museum. While he was still a student, Skottsberg was the official botanist on the 1901–04 Swedish Antarctic Expedition led by Otto Nordenskjöld (1869–1928), and he led the 1907–09 Swedish Magellan Expedition to southern South America. He returned to Juan Fernández in 1916–17 and again in 1954–55, and he made four visits to Hawaii to study the flora. Skottsberg also traveled widely in Europe.

In 1915 the city of Göteborg commissioned Skottsberg to design a botanical garden. Work was completed in 1919, and Skottsberg became its director, a position he held for the next 29 years. From 1924 to 1947 Skottsberg was secretary of the Royal Society of Science and Letters of Göteborg. He was secretary of the International Commission for Preservation of Wildlife in the Pacific from 1929 to 1949. In 1949 he became president of the Royal Swedish Academy, and in 1950 he was elected president of the Seventh International Botanical Congress and also a foreign member of the Royal Society of London. Skottsberg died at Göteborg on June 14, 1963.

Plant Cultivation

There is an ancient legend describing the supposed origin of the game of chess. The king of a prosperous country—different versions of the story locate it in various parts of the world—summoned his wisest adviser and asked him to devise a board game in which success would depend entirely on the skill of the players. This game, the king hoped, would prove popular among his people, who were growing addicted to gambling and games of chance. The wise man devised just such a game to be played on a board divided into 64 squares. The king was delighted and asked his adviser to name his own reward. The wise man replied that his needs were modest. All he asked was one grain of wheat for the first square on the board, two for the second square, four for the third, and so on, doubling the number of grains for each square. The king readily agreed and sent his officials away to count out what he imagined was a very small payment. When they began to do the math, however, they quickly discovered that the wise man had lived up to his reputation by asking for 2 + 4 + 8 + 16 + 32 + 64 + 128 + 256 . . . grains and when they came to the end they must start over taking a different square as the first, so the final score had to be multiplied by 64. It amounted to a total of more than 18 quintillion (18×10^{18}) grains of wheat. Not only was this far more wheat than all the granaries in the world could hold when full, it was far more than his country was capable of producing within the adviser's lifetime, even if he lived to a very old age.

The story has three lessons. The first is that chess has been a popular game of skill for a very long time. The second is that those who make rash but seemingly innocuous promises should first learn to check the math. The third is that wheat is of vital importance.

This chapter tells of the origin of a few of the most important food crops. These include wheat, rice, and corn, which are the staple foods over much of the world. Cotton has also been cultivated since very ancient times, and the chapter also tells as much as is known of its origin. The plants yielding these crops were domesticated long before writing was invented. Unraveling their history is one of the most fascinating—and most challenging—branches of science, where archaeology, botany, and genetics intersect.

Tea and coffee are stimulants, and their origins are associated with legends, which the chapter recounts. Finally, the story tells of the way certain plants were transported across the globe, and how some triggered the emergence of large and profitable industries in lands far from where they had originated.

THE ORIGINS OF AGRICULTURE

Life in a modern city or even a large town would be impossible were it not for the farmers who grow crops and the distribution and retailing systems that bring the food to the stores. It is agriculture that makes urban life possible, but it makes no sense to suggest that farming began in order to make it possible for people to live in cities. How could people have had any concept of cities in a world where none existed?

Originally, humans obtained their food by hunting, scavenging, fishing, and gathering edible plant materials such as fruits, roots, and leaves. In some places people manipulated their environment to encourage it to produce more food. They set fires to drive game into ambushes, and the fires also cleared away wilted and inedible plants, fertilized the soil with the ash, and produced new plant growth. In some places they weeded food-producing areas to remove inedible plants that competed with those yielding tasty, satisfying foods. People knew, of course, that plants grow from seeds, and there were—and are still, for instance in tropical South America—tribes who cleared patches of forest to make gardens in which they cultivated vegetables.

This was not agriculture, however. Agriculture is the cultivation of domesticated crop plants, especially the cereals that provide the basic ingredients in most diets. No one knows just why people gradually changed from hunting and gathering to farming. There have been many suggestions.

Some have proposed that the invention of farming was an inevitable consequence of cultural advance. Once a few people had discovered how crops might be cultivated, the superiority of farming over living on wild foods became evident. Farming would allow nomadic people to settle in one place and enjoy a higher standard of living with enhanced food security. They would have time to devote to art, craft, music, and intellectual debate. This seems unlikely, however. Most contemporary hunter-gatherers find their lives quite satisfactory. They do not necessarily have to devote more time to obtaining food than they would if they grew it, and most seem to enjoy plenty of leisure. They have a detailed knowledge of the plants and animals on which they depend and in some cases hunter-gatherers live close to agricultural communities and trade with them, without being tempted to change their way of life.

Perhaps, then, some combination of demographic and environmental change made hunting and gathering less dependable. Maybe a prolonged period of favorable conditions allowed populations to increase, and when conditions reverted to their former state the natural environment could no longer produce enough food for everyone and people were forced to start farming. That might be what happened, but there is a difficulty: In its early stages there is no guarantee that farming will increase food availability.

A more likely explanation is that no one set out to invent farming. It happened almost by itself when, over many generations, people selected those individual plants that best suited their needs. Spilled seeds from the chosen plants would have germinated, people would then have started to combine cultivation of some crops with the gathering of others, and in time cultivation would have come to provide most of the community's food. As people learned to grow crops, their selection of individual plants altered the characteristics of the crop species until farming communities and their crops were inseparable. Whatever the reasons for it, crop cultivation arose independently in many parts of the world, and archaeologists know approximately when this happened. The following table lists the regions where farming began, and when.

ORIGIN OF FARMING	
REGION OF THE WORLD	WHEN FARMING BEGAN (YEARS AGO)
Southwest Asia	11,500
Northern China	More than 9,000
Southern China	8,000
Central Mexico	5,750
Andes of Peru	5,250
Eastern North America	5,250
West Africa	4,500
Papua New Guinea*	9,000
Amazonia*	9,000
*It is uncertain when cultivation commenced here.	

Although scientists are uncertain when cultivation began in Amazonia, there is overwhelming evidence that for centuries prior to the arrival of Europeans much of what is now lowland rain forest was farmed. William Denevan, a professor emeritus at the University of Wisconsin, Madison, devoted many years to the study of this region and has written extensively on the farming that was once practiced there.

There are the remains of what were once fields. Some fields were raised to improve drainage. Others were worked into ridges and furrows; crops would have been sown along the ridges. There were fields worked into scattered raised areas and fields with ditches to remove surplus water. The soils were also changed. The Amazonian farmers treated them with charcoal and possibly composted wastes. These treatments blackened the soils, so they are quite unlike other soils in the region. These relics of former farms cover a large area, and it is likely that until a few centuries ago farms covered a substantial part of what is now forest. Far from having existed for thousands or even millions of years, these Amazonian forests are no more than about 500 years old. The farmers would have been able to support villages with a combined population in the hundreds of thousands.

THE STORY OF WHEAT

Wheat is a type of grass that grows wild in the mountains of southeastern Turkey, and it was there, about 11,000 or possibly as much as 12,500 years ago, that people began to cultivate it. They had been collecting wheat grain from time immemorial, preferring einkorn (*Triticum monococcum*) and emmer (*T. turgidum*) wheats, but they faced a problem. Dense though the stands of these wheats were—and still are—it was impossible to avoid losing much of the grain. In all cereals, the grains are the plants' seeds, and when the ears are ripe wild plants shed the seeds spontaneously, so they fall to the ground where they are able to germinate. Once on the ground, however, most of the wild grain was lost to the human communities that depended on it. So people harvested the crop before it shed its seeds and threshed the plants to remove the seeds in a place where they could be swept up. It meant that many of the ears were not quite ripe, but at least the grains were not lost.

Cereal grains are attached to the shaft of the ear, called the *rachis*. When the grains are ripe the rachis becomes brittle and breaks, allowing the grains to fall. Among the wild plants that people collected there were some in which the grains were ripe but the rachis had failed to break. After a time people began to plant some of the seeds they had collected and harvested a crop they had cultivated. This was the beginning of domestication, and it had several consequences that *archaeobotanists*—botanists who study plant remains recovered from archaeological sites—are able to recognize.

When a cereal seed is shed naturally, it bears a neat, round scar at the point where it detached from the rachis. If the seed was harvested while still attached to the plant and separated by threshing before the rachis became brittle, fracturing of the rachis produces a rougher scar. This allows archaeobotanists to distinguish wild from domesticated grains.

Wild cereal seeds possess hooks that curl and straighten in response to being wetted and dried. When they reach the ground this movement makes them drag themselves beneath the soil surface. This hides them from seed-eating birds and mammals and provides them with suitable conditions for germination. Seeds of domesticated cereals lack these hooks.

A cereal seed contains *endosperm*—a store of food to sustain the young plant until it is able to fend for itself. Wild seeds germinate very close to the soil surface, so the emerging shoots have only a short distance to grow before they are exposed to sunlight and commence photosynthesis. Consequently, the seeds require only a small amount of endosperm. Farmers sow the seeds much deeper in the soil. This

© Infobase Publishing

A: The earliest cultivated wheat was einkorn (*Triticum monococcum*). B: Modern bread wheat (*T. aestivum*) in its awned (right) and C: unawned forms

means that only the largest seeds, with the most endosperm, are likely to survive. Domestication produces larger grains.

Early farmers were inadvertently selecting the traits that over many generations led to the cultivated wheat varieties grown today. They selected plants that produced large seeds and, more important, plants with tough rachises that did not shed their seeds as soon as they ripened. A single genetic mutation makes the difference between a brittle rachis and a tough one. Natural selection favors the brittle rachis, because it allows the plant to release its seeds. Artificial selection by farmers favored the tough rachis, because it allowed the grain to be harvested. The farmers won, but in doing so they produced cereals that were entirely dependent on humans because they were unable to propagate efficiently. The following illustration compares einkorn wheat with two varieties of modern bread wheat (*T. aestivum*), one with *awns*—long bristles that protrude from the grains—and one without.

The transformation of wild wheats into modern domesticated wheats was far from straightforward, and it involved several wheat species and related grasses. Nor did it happen quickly. For a long time farmers would have had no way to prevent wild cereals from growing among their cultivated crops and hybridizing with them. Perhaps the wild plants periodically overwhelmed the cultivated ones. When the climate was favorable and the wild cereal stands flourished, communities might have abandoned farming and all the hard work it entailed, preferring to go out with their stone-bladed sickles and gather their food until weather conditions deteriorated and they resumed cultivation. After such a vast expanse of time, no one can tell. What archaeobotanists do know, however, is that it took thousands of years to domesticate wheat. From the time that grains with a tough rachis first appeared, it was more than 3,000 years before this mutation came to predominate, showing that the domestication process was complete.

THE STORY OF RICE

Rice (*Oryza sativa*) is also a member of the grass family (Poaceae). Today it is the staple food for about half the world's population, with two major varieties, *O. sativa indica* and *O. sativa japonica*. Its many tiny flowers, known collectively as an *inflorescence*, have a complex structure with many branches. This is called a *panicle,* and as the

flowers develop into seed grains these also form panicles. The following illustration shows a rice panicle.

Rice was first cultivated in the valley of the Yangtze River in
China. As with wheat, rice domestication involved the mutation that
substituted a tough rachis for a brittle one and, also as with wheat,
archaeobotanists can tell by the rachis scar left on the grain whether
it came from a wild or domesticated plant. As well as the evidence
from the grains themselves, rice cultivation attracts a distinctive
suite of weeds. Although rice can be grown as a field crop on dry
land, it is most productive when the young seedlings are planted into
shallow ponds called *paddies* to complete their growth. The paddies
are then drained when it is time for the crop to ripen. Plants such as
sedges and rushes commonly occur as weeds in rice paddies, and the
presence of these plants at archaeological sites alongside rice grains
is clear evidence of paddy cultivation.

© Infobase Publishing

Rice grains develop on
a complex, branching
arrangement of stems called
a panicle.

Rice was first domesticated along the Yangtze Valley in China to the south of Shanghai. Research at Tian Luo Shan shows that the process was not completed more than 2,000 years after it began at Kuahuqiao.

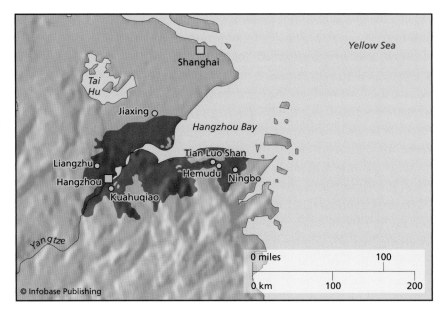

There is archaeobotanical evidence of rice cultivation between 8,000 and 7,700 years ago at Kuahuqiao in Zhejiang Province, about 20 miles (32 km) upstream from the head of the Yangtze estuary. Investigations at Tian Luo Shan led by Dorian Fuller of University College London have shown that the domestication process had not been completed by about 6,600 years ago. This is further evidence that crop domestication is a very slow process. The above map shows the location of these archaeological sites.

After the plant had been domesticated in China, rice cultivation spread eastward to Korea and Japan and westward into India. Farmers were growing rice in northeastern India and Myanmar (Burma) by about 4,000 years ago, and it reached southern India by about 1,400 years ago.

THE STORY OF CORN

Corn, also called maize (*Zea mays mays*), is a crop plant that originated in America. It is one of the approximately 10,000 members of the grass family (Poaceae), but it differs from the other cereal grasses in the extent to which the process of domestication altered it. These differences are so great that at one time botanists believed that the wild ancestor of maize had long been extinct. With the other cereals—wheat, barley, rye, oats, and rice—there are wild plants that closely resemble the domesticated crop plants. Indeed, wild oats (*Avena fatua*), which closely resemble cultivated oats (*A. sativa*) in appearance, are a troublesome weed of cereal crops. Corn is different. Its wild ancestor still thrives, but it looks so different from cultivated corn that botanists formerly placed the two plants in different genera.

Corn is the domesticated form of teosinte, a name derived from *teocintli*, which means "grain of the gods" in the language of the Nahuátl people of Central America. There are five species of teosinte grasses found in Mexico and Guatemala, some of them *annual* and others *perennial*—an annual plant completes its life cycle, from germination to releasing seeds, in a single year; a perennial plant lives for more than two years and flowers annually after an initial period during which it may not flower at all. Corn is descended from one of the annual species, *Zea mays parviglumis*.

Teosinte produces ears with between five and 12 kernels. The kernels are small and enclosed within hardened coatings that protect them from animals that might eat them. When the kernels ripen, the plant sheds them and the protective coatings disappear. Domestication began about 9,000 years ago in the Balsas River Valley in southern Mexico. People began cultivating teosinte, selecting the plants with the biggest ears, tough rachises, and tastiest kernels, and in time a number of mutations occurred that made the mutated plants still more attractive, so those were the ones that were grown. The wild and cultivated forms continued to interbreed—and cultivated corn and wild teosinte are still capable of interbreeding—but the changes continued to accumulate. Modern corn has an ear with several hundred kernels that lack any protective outer coating or any means of releasing and dispersing the seeds. The crop is wholly dependent on humans for its survival.

From Mexico, corn cultivation spread by several routes into North and South America. One route led from the Mexican uplands across the highlands of Panama and into the Andes on the western side of South America. Another began in the Central American lowlands, spread into the coastal regions of northeastern South America, and spread from there into the continental interior along the river valleys. People were growing corn in northern South America about

4,500 years ago, but it was not until about 1,500 years ago that farming became established in Chile. Farming also expanded northward, reaching the southwestern region of North America between about 2,500 and 2,000 years ago. Once established there, the cultivation of corn provided the dietary and economic stability that led to the development of the Pueblo cultures of the Southwest.

THE STORY OF COTTON

No one knows who were the first people to wear clothes made from cotton or where they lived. Archaeologists have found traces of cotton fabrics dated at about 2300 B.C.E. in the ruins of Mohenjo-Daro. At that time the two cities of Mohenjo-Daro and Harappa, in the Indus Valley along what is now the border between Pakistan and India, were just emerging as the centers of a civilization that lasted until about 1700 B.C.E., when the climate became drier, crops failed, and the cities and other settlements were abandoned. People were also wearing cotton garments between 3500 and 2300 B.C.E. in the Tehuacán Valley of Mexico. The Mexican textiles were made from fully domesticated cotton (*Gossypium hirsutum*), suggesting an earlier history of domestication that may have taken place in South America, but based there on a different species, *G. barbadense.* It was being used in 3600 B.C.E. in the northern part of the Atacama Desert in Chile, and in 2500 B.C.E. along the coasts of Ecuador and Peru. At that time the cotton was still at an early stage in its domestication. Mexican cotton (*G. hirsutum*) still grows wild along the coasts of Central America and southern North America, but wild South American cotton (*G. barbardense*) is now found only along the coast of Ecuador.

The cotton plant is either a low-growing herb, a shrub about two feet (60 cm) tall, or a small tree up to six feet (1.8 m) tall. It may be an annual, *biennial,* or perennial—a biennial plant is one that completes its life cycle in two years. Modern cotton growers treat the plant as an annual, planting it anew each year. That is a consequence of domestication, because growers repeatedly selected annual plants.

Cotton is useful because it yields several products. Its seeds develop inside a covering of hairs that are 3,000 times longer than they are wide, mixed with much shorter hairs called fuzz. The long hairs are spun into threads, and one pound of raw cotton contains about 90 million individual fibers (200 million per kilogram). The fuzz is made into felt, cotton wool, and other products. Cottonseeds are pressed for their oil, the seeds can also be made into a high-protein flour, and seedcake made from pressed cottonseeds is a nutritious cattle feed.

Cotton is unusual among crop plants in that it was domesticated independently twice in Central and South America, once in Africa, and once in Asia. There are 39 species of *Gossypium* that grow naturally in warm-temperate and tropical regions. Of these, four species have been domesticated.

African or West Asian cotton (*G. herbaceum*) is a perennial shrub found in savanna and semidesert environments in Arabia and in Africa south of the Sahara. It was domesticated in either Ethiopia or southern Arabia, and knowledge of its cultivation spread eastward into Persia (Iran), Afghanistan, Turkestan, and China, where it was being grown in about 600 C.E., and northward into Turkey, Ukraine, and southern Europe.

Asian cotton (*G. arboreum*) is the species that was domesticated in the Indus Valley. Some varieties of Asian cotton are annual plants and others are perennial shrubs. One of the perennial varieties was introduced to East Africa, and about 2,000 years ago the Meroe people of Nubia were growing it. In ancient times Nubia was a region to the south of Egypt, extending westward to the edge of the Libyan Desert, southward about to Khartoum, and eastward to the Red Sea. The cultivation of Asian cotton spread westward from Nubia, eventually reaching Nigeria. In the ninth century the city of Kano, Nigeria, was a center for cotton textile production.

South American cotton (*G. barbadense*) was fully domesticated by 1000 B.C.E., and its cultivation spread throughout much of South America and eastward to the Caribbean islands. Christopher Columbus encountered it in Barbados, which is how it acquired the specific name *barbadense.* This was the species that came to be grown on the slave plantations of the British West Indies. By the 1650s Barbados was exporting cotton, and in about 1670 cotton planters brought the crop from Barbados to the British North American colonies, founding the North American cotton industry.

Mexican cotton (*G. hirsutum*) was being grown extensively in many parts of Central America by the time the first Spanish explorers arrived in the early 16th century. The Maya and Aztec civilizations made great use of it, and by the first century C.E. Mexican cotton was also being grown in Arizona. Spanish colonists sent

Mexican cotton to Europe, and in time its superior qualities allowed it to displace the other cultivated species. Most commercial cotton is now *G. hirsutum.*

CAPTAIN BLIGH, HMS *BOUNTY,* AND THE BREADFRUIT TREES

In the late 18th century Sir Joseph Banks (1743–1820) was president of the Royal Society. For a time Banks had been the unofficial director of the Royal Botanic Gardens at Kew (see "Sir Joseph Banks, Unofficial Director at Kew" on pages 70–72). He wielded considerable influence in political and scientific circles.

Britain was establishing colonies in many parts of the Tropics, and there was much interest among politicians and administrators in identifying potentially useful plants and transplanting them to places where they might be cultivated profitably. Breadfruit was one such plant. This is an unusual fruit in that its principal nutrient is starch—the fruit is about 25 percent carbohydrate and 70 percent water. It is most often eaten roasted, baked, boiled, or fried, but it can also be made into other dishes. The important point is that it is both nutritious and productive—under ideal growing conditions in southern India a tree can produce up to 200 fruits every year, although it is less productive elsewhere. The British saw it as an easily cultivable food for the slaves on their West Indian plantations. The illustration opposite shows two ripe breadfruits.

The breadfruit tree (*Artocarpus altilis*) grows to a height of about 66 feet (20 m). It is monoecious, the male flowers appearing first, and pollinated by fruit bats (Megachiroptera), but trees are also propagated by planting the suckers that grow naturally from the roots. The rough-skinned fruits are about the size of grapefruits. The tree occurs naturally in the Malay Peninsula and the islands of the South Pacific, where Polynesians spread it widely by carrying cuttings on long voyages and planting them when they reached their destinations.

Banks gave his support to a government scheme to gather breadfruit tree cuttings in Tahiti and transport them to the West Indies. The plan was agreed, the ship chosen for the task was the *Bounty,* and, on Banks's recommendation, the naval officer given command of the *Bounty* was Lieutenant William Bligh (1754–1817). Funding for the enterprise came from a prize awarded by the Royal Society

Starch is the principal nutrient in breadfruit (*Artocarpus altilis*). That is why the British government thought it would be a useful crop to grow in the West Indies. This breadfruit tree is in Jamaica. *(Fred McConnaughey/Photo Researchers, Inc.)*

of Arts. The *Bounty* was a collier that had been converted into an armed vessel by fitting 10 swivel guns—small cannon mounted on a stand so they could be turned—and four four-pounder (2-kg) cannons. She had three masts and was fully rigged, but she was small, approximately 91 feet (27.7 m) long and 24 feet (7.3 m) wide. The ship carried a complement of 44 officers and men. There was a garden to supply fresh vegetables and pots secured along both sides for the tree seedlings. It is important for what followed that no marines sailed on the *Bounty*. Most warships carried marines, because the standard method for fighting at sea involved boarding the enemy vessel and overpowering its crew in hand-to-hand combat—a job for soldiers, not seamen.

The *Bounty* set sail on December 23, 1787, bound for Tahiti. The ship crossed the Atlantic, but after spending a month trying to round Cape Horn in adverse winds, she was forced to turn back. She passed round the Cape of Good Hope and sailed across the Indian Ocean into the Pacific. During the voyage Bligh—who was then 33 years old—demoted the ship's sailing master John Fryer (1753–1817), whom he found unsatisfactory, and promoted his protégé Fletcher Christian (1764–93) to replace him, with the rank of acting lieutenant. Despite fictional portrayals of him, Bligh was not a harsh or abusive commander and when he awarded punishments they were more lenient than those that sailors could expect on most ships. Bligh made sure his men had time for exercise, ensured they ate wholesome food including lime juice and sauerkraut as reliable sources of vitamin C, and he demanded that his men bathe and wash their clothes regularly.

They reached Tahiti—then called Otaheite—on October 26, 1788, and spent five months there gathering and potting 1,015 breadfruit saplings they had grown from seed. During this time Bligh permitted his crew to live ashore, and Christian married a Tahitian woman, Maimiti. On April 4, 1789, the *Bounty* left Tahiti. The mutiny, led by Christian, broke out on April 28. The reasons are obscure, but Christian and his followers may have been motivated by a wish to return to the easy life they had enjoyed ashore. No one was injured in the mutiny, and had the ship carried marines loyal to the captain probably it would never have taken place. The majority of the crew took no part in the mutiny, but they did nothing to prevent the mutineers from binding Bligh and ordering him, together with 18 of his crew, into a 23-foot (7-m), long launch with four cutlasses, food and water, a sextant, and a pocket watch. The *Bounty* then sailed away from them.

Bligh and his companions sailed first to Tofua Island in the Tonga group to take on supplies, but they came under attack from hostile islanders and one member of Bligh's crew was killed. They sailed away, not daring to call at other possibly

Captain William Bligh (1754–1817) was the commander of HMS *Bounty,* scene of the famous mutiny. This drawing, made by G. Dance on January 1, 1805, shows Bligh at the age of 50. *(Hulton Archive/Getty Images)*

hostile islands, and Bligh navigated them to the island of Timor, off the tip of the Malay Peninsula, where there were European settlements, a distance of 3,618 nautical miles (6,700 km). The journey took them 47 days, and no other men were lost. It was a magnificent feat of seamanship. Bligh returned to London in March 1790. He was court-martialed for losing the *Bounty*, but acquitted, and from 1791 to 1793 he was master and commander of HMS *Providence,* which, accompanied by HMS *Assistance,* completed the task of transporting breadfruit trees from Tahiti to the West Indies. Bligh remained in the navy until 1805, when he was appointed governor of New South Wales, but he was deposed in 1808 when colonists rebelled against his attempts to stamp out corruption. The illustration opposite shows him at about the time of his appointment to the governorship. In 1814 Bligh was promoted to the rank of vice admiral.

William Bligh was born on September 9, 1754, in the village of St. Tudy, near Bodmin in Cornwall. Promotion in the Royal Navy depended on length of service, and so it was common for parents to "sign on" their sons at a young age. Bligh officially joined the navy at the age of seven. He died in London on December 7, 1817.

TEA, AND HOW BODHIDHARMA STAYED AWAKE

The founder of Zen Buddhism was a sage from southern India whose Sanskrit name was Bodhidharma, which means "teachings of the Buddha." In China he is called Damo, and in Japan he is Daruma. The son of a Tamil king, Bodhidharma lived in the fifth or sixth century C.E. and became a Buddhist monk. He introduced his school of Buddhism during his travels in China, where it is known as Ch'an Buddhism. This version of Buddhism entered Japan during the Kamakura era (1185–1333); Zen is its Japanese name.

There are countless stories about Bodhidharma. One tells that the sage kept falling asleep while meditating. This so enraged him that he cut off his own eyelids to ensure that he would never sleep again. He threw his eyelids onto the ground and where they landed the first tea plants grew. An alternative legend holds that he carried tea plants with him when he walked from India to China.

Buddhist practice centers on meditation, and Zen monks spend hours on end meditating. One monk spends each session observing the others and striking any who doze off with a stick. In some monasteries the monks also drink green tea at intervals to help keep them awake. The new teaching quickly became popular among the samurai (warrior) class, and the serving of tea led to the development of the tea ceremony.

After his death, some people reported seeing Bodhidharma walking in the direction of India carrying one shoe in his hand. When his grave was opened it was found to contain the shoe he had left behind.

COFFEE, AND KALDI'S GOATS

Coffee originated in Ethiopia. That is where the plants (*Coffea arabica* and other species) grow naturally, and it is where they were first cultivated. Ethiopians were the first people to drink coffee.

According to legend, it all began with a goatherd called Kaldi who lived in the ninth century C.E., although it was not Kaldi but his goats who made the discovery. One version of the story states that each morning Kaldi would turn his goats loose to wander in the hills and each evening they would return home of their own accord. Except that one evening they failed to return, and Kaldi grew worried. He searched high and low, and the following morning he found them jumping and frisking around a small patch of bushes with red berries and dark, shiny leaves. He assumed the bushes were responsible for the odd way the goats were behaving, so he tried the berries for himself and very soon he, too, was jumping around with boundless energy. The alternative version says that Kaldi did not turn the goats loose but remained with them all day, and one afternoon he noticed them behaving strangely some distance away.

Kaldi gathered some of the berries and either he met a wise monk on his way home and gave the berries to him, or he arrived home, gave the berries to his wife, and told her what had happened. Convinced that they were a gift from god, she took them to a nearby monastery. One way or another, the berries came into the possession of the monks. Some say that the monks denounced them as products of the devil and threw them onto the fire; others that the monks studied the berries closely and experimented with them. As they roasted, the berries released a delicious aroma, attracting other monks. The monks collected the roasted berries and steeped them in boiling water to preserve them, then experimented by drinking the resulting brew. They discovered that the fragrant drink helped them stay awake during night prayers.

Knowledge of the coffee plant and its properties spread to Egypt, Yemen, and Arabia, where it quickly became popular. The Arabs produced and marketed coffee commercially, and it was from them that it reached Europe. Today Brazil is the world's leading coffee producer, with an annual output of about 1.4 million tons (1.3 million tonnes). Global production is about 7.7 million tons (7 million tonnes).

HOW BRAZIL ACQUIRED ITS NAME

In the 15th and 16th centuries, wealthy Europeans dressed in rich fabrics dyed with bright colors, while ordinary folk were clad in dull grays and browns. Red was especially popular, not least because the dye, derived from the wood of the sappanwood tree (*Caesalpinia sappan*), was so expensive that sporting a red coat or gown was an ostentatious display of wealth, the equivalent of driving a Cadillac or Rolls Royce today. The dye, with the color of burning coals, was imported from southern Asia and reached Europe in powder form. The Portuguese called the tree *pau-brasil, pau* meaning "wood" and *brasil* meaning "ember." In English it was known as brazilwood.

On April 22, 1500, Portuguese sailors reached the coast of South America. When they went ashore they saw trees that had dense, orange-red wood and soon discovered that the wood yielded a red dye. These pau-brasil trees (*Caesalpinia echinata*), also known as pernambuco and every bit as good as sappanwood, grew abundantly along the coast and beside river courses extending far inland. The explorers sent samples back to Europe, and within a few years the trees were being felled and shipped across the ocean in large quantities. It proved such a highly profitable industry that ships laden with pau-brasil timber were at risk from pirates, and in 1555 a French expedition of two ships and 600 men led by Nicolas Durand de Villegaignon (1510–71), vice-admiral of Britany, attempted to establish a French colony they called France Antarctique at Rio de Janeiro, partly in order to gain access to this priceless timber. The territory from which pau-brasil was gathered became known in Portuguese as Brasil and in other languages as Brazil.

Brazilwood is still used as the source of the dye brazilin, and the timber is also used to make bows for stringed instruments. The brazilwood industry collapsed in the 18th century because so many of the trees had been felled that they were becoming rare. The species is now classed as endangered.

SIR HANS SLOANE, MILK CHOCOLATE, AND THE BRITISH MUSEUM

Cocoa is the most popular alternative drink to coffee and tea. It is made from the berries of a small tree (*Theobroma cacao*) that is native to tropical America and the islands of the Caribbean including Jamaica, which is the source of the first cocoa to reach Europe in 1698. The berries develop inside pods, surrounded by mucilage. The

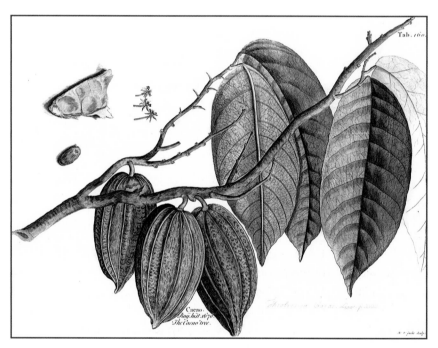

Cocoa pods (*Theobroma cacao*) contain the beans that, after processing, yield cocoa butter and, after most of this has been removed, cocoa powder. This drawing of the pods is from *A Voyage to the Islands Madera, Barbados, Nieves, S. Christophers and Jamaica, with the Natural History of the Herbs and Trees, Four-footed Beasts, Fishes, Birds, Insects, Reptiles, &c. Of the last of those ISLANDS* by Hans Sloane (1660–1753), in two volumes published in 1707 and 1725. *(The British Museum)*

beans and mucilage are scooped from the pods and fermented. After fermentation the beans are a dull red. They are dried and are then ready to be exported. The dried beans contain 50 to 57 percent of a fat called cocoa butter. Most of the fat is removed before the beans are ground to make cocoa powder, but when the powder is used to make chocolate some of the butter is restored. The illustration opposite shows cocoa pods.

The drawing first appeared in a two-volume work on natural history, published in 1707 and 1725, with the title *A Voyage to the Islands Madera, Barbados, Nieves, S. Christophers and Jamaica, with the Natural History of the Herbs and Trees, Four-footed Beasts, Fishes, Birds, Insects, Reptiles, &c. Of the last of those ISLANDS.* The author was Hans Sloane, an Irish-born naturalist and physician, who had toured the region he described and had spent 15 months in Jamaica. It was in Jamaica that Sloane first encountered cocoa. Local people mixed it with water and used it as a medicine, but Sloane described it as "nauseous." Nevertheless, he persisted with it and found he could improve its flavor by mixing it with milk to make chocolate. He devised a recipe for this and publicized it on his return to England. After a time people were able to purchase chocolate from apothecaries. In the 19th century Cadbury Brothers began manufacturing chocolate using Sloane's recipe, but as a luxury food rather than a medicine.

Hans Sloane was born on April 16, 1660, in Killyleagh, County Down, Ireland. Even as a child he was an avid collector of curiosities, including coins, medals, seals, drawings, and natural history specimens. He studied chemistry at Apothecaries Hall in London from 1679, taking a particular interest in medical botany, which he studied at the Chelsea Physic Garden. He became a friend of John Ray (see "John Ray and His Encyclopedia of Plant Life" on pages 16–18). In 1683 he toured France, studying anatomy, medicine, and botany, and received his degree of doctor of physics (pharmacy) later the same year. While in France he became friendly with Joseph Pitton de Tournefort (see "Joseph Pitton de Tournefort and the Grouping of Plants" on pages 80–83). Sloane returned to England in 1685 and was elected a fellow of the recently formed Royal Society. In 1687 he was made a fellow of the Royal College of Physicians. It was at that time that he was invited to travel to Jamaica as personal physician to the new governor.

Sir Hans Sloane (1660–1753) was an Irish-born naturalist and physician who introduced cocoa to Europe and left a large collection of books and curiosities to the nation. The British government founded the British Museum to house it. *(Science Photo Library)*

Following his return, Sloane set up a medical practice in London in 1695. The following year he was made physician extraordinary to Queen Anne and in 1716 to George I. In 1727 he was made physician in ordinary to George II. His practice was obviously successful, but he continued to find time for botany and for adding to his collections. He was made a baronet in 1716. He became president of the Royal College of Physicians in 1719 and of the Royal Society in 1727. The portrait at left shows Sloane at the height of his success.

Sir Hans Sloane retired in 1741. By that time his collection was vast and of great value, and it included collections of natural history specimens he had received from other eminent collectors. Sloane died on January 11, 1753, leaving his collection to the nation on condition that Parliament pay £20,000 to his executors. This was far less than the collection was worth, and Parliament accepted. The collection, together with a number of other collections, was opened to the public in 1759 as the British Museum. Much of the natural history collection later formed the basis of the Natural History Museum.

HOW RUBBER MOVED TO ASIA

Natural rubber is made from the *latex* exuded by a tree, *Hevea brasiliensis.* Latex is a milky fluid produced by some herbs and trees that may carry nutrients and may also help the plant to heal wounds. The rubber tree grows naturally in tropical South America, where people were using rubber long before the arrival of the first Europeans. In 1525 a Spanish priest, Padre d'Anghieria, wrote that he had seen people in Mexico playing with balls that bounced. It was not until 1735, however, that the French explorer, geographer, and mathematician Charles de La Condamine (1701–74) made the first scientific study of rubber during a visit to Peru. The first use Europeans found for this substance was as an eraser. In England it became known as India rubber. Then its waterproof qualities were recognized. Native South Americans used rubber to make waterproof shoes, and in the early decades of the 19th century North American manufacturers began producing waterproof garments and footwear containing rubber.

Rubber becomes brittle when it is very cold, and plastic when it is warm. These disadvantages restricted its use until 1840, when Charles Goodyear (1800–60) discovered vulcanization, an industrial process that stabilized rubber, allowing it to be used over a much wider temperature range. American and European factories began using rubber to make flexible tubing, pneumatic tires—first invented in 1845, but then forgotten and reinvented in 1869—and soft toys. The rubber industry thrived, and demand for the raw material triggered a boom in Brazil, where the latex was collected from trees growing wild in the rain forests of Amazonas. Amazonas prospered, and thousands of immigrants moved into the region to collect and sell latex. By about 1870 Brazil dominated world rubber production.

The Brazilian monopoly suffered a fatal blow in 1876. In that year the English explorer Sir Henry Wickham (1800–67) gathered about 70,000 seeds from wild rubber trees in the forest close to the city of Santarém, in the state of Pará. Wickham smuggled the seeds out of Brazil and took them to Kew Gardens, London, where they were sown. Many of them germinated, and 3,000 seedlings were sent from London to Ceylon (now Sri Lanka). In 1877, 22 rubber plants were sent from Ceylon to the Singapore Botanic Gardens. The trees were growing there when in 1888 Sir Henry Nicholas Ridley (1855–1956) arrived as the gardens' first scientific director. Ridley spent years studying the trees, and in 1895 he discovered a technique for tapping the latex without seriously harming the tree. That made it practicable to cultivate the trees commercially. In 1890 Ridley exhibited the first cultivated rubber trees, and in 1896 the first rubber plantations were established in Malaysia. Most of the trees were grown from Ridley's seeds. Growers went on to produce hardier, disease-resistant varieties, and large rubber plantations were developed in Ceylon and Singapore as well as Malaysia.

In rubber plantations the trees were planted 13 feet (4 m) apart. In the forests of Amazonia, several miles often separated one wild rubber tree from the next, and the traditional method for tapping the latex damaged the trees. The Brazilian producers could not compete with the more efficient Asian plantations, but the Brazilian government was reluctant to introduce plantation production because it relied on the people working the wild trees to maintain its territorial claim over the sparsely populated Amazonas region. The Brazilian boom ended, and the migrant workers drifted away from the forest. The motor manufacturer Henry Ford (1863–1947) sought to revive the Brazilian rubber industry in the 1920s. His company planted more than 70 million rubber trees in Pará State, aiming to produce 300,000 tons (270,000 tonnes) of rubber a year. Local people dubbed the plantations Fordlândia, but the harsh Amazonian environment caused the project to fail and the plantations were abandoned, leaving the Asian plantations producing more than 90 percent of the world's natural rubber.

The industry changed once more during World War II, when the Japanese took control of the Asian plantations. In the United States rubber products were recycled in the largest recycling operation there has ever been. It was forbidden to use rubber for anything that was not directly linked to the war effort, the speed limit on all highways was reduced to 35 MPH (56 km/h), and scientists were directed to develop alternatives to rubber. A synthetic rubber industry came into existence. Today there are about 20 grades of synthetic rubber, made from crude oil, and together they supply about 75 percent of the global rubber industry.

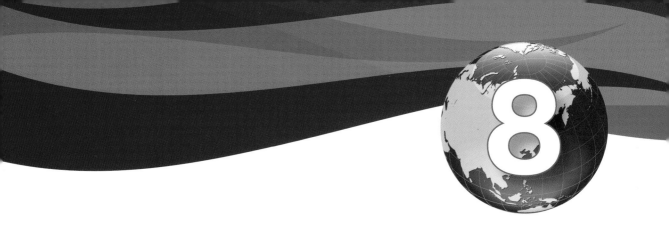

Evolution of Plants

By the early years of the 19th century, natural scientists had come to recognize that the plants they observed and collected differed from those that their geologist colleagues found as impressions in coal and slates. This clearly implied that certain plants had become extinct in the distant past and others had taken their places. The most logical explanation for this was that plants had changed over long periods. In a word, they had evolved.

This chapter tells of some of the steps by which scientists traced the evolution of plants, beginning with the work of the man who is often described as the father of *paleobotany*—the scientific study of plant fossils and other traces and remains in order to reconstruct past environments and the evolutionary history of plants. It describes the development of the theory of evolution by means of natural selection and explanations for the discontinuous distribution of certain plants.

Cultivated plants have also evolved as a consequence of their domestication. The chapter ends by describing the work and life of the Russian biologist who devised a method for identifying the regions of the world where modern crop plants were first cultivated.

ADOLPHE-THÉODORE BRONGNIART, FATHER OF PALEOBOTANY

In 1822 a 21-year-old French botanist published a paper on the distribution and classification of fossil plants. This established the career path the young man would follow and that culminated with his most important work. This began with *Prodrome d'une Histoire des Végétaux Fossiles* (Introduction to a history of fossil plants), published in 1828, followed by the two-volume *Histoire des Végétaux Fossiles* published in 1828 and 1837. The author was Adolphe-Théodore Brongniart (1801–76), and it was this work that gave him the reputation of being the father of paleobotany.

Brongniart's father, Alexandre Brongniart (1770–1847), was a chemist, mineralogist, and zoologist who taught at the École des Mines (mining school) in Paris. For several years, Brongniart senior collaborated with the zoologist Georges Cuvier (1769–1832) in a study of the geology of the Paris basin, a region of sedimentary rocks deposited 65.5 to 5.3 million years ago during the Tertiary sub-era. Cuvier, one of the most eminent French scientists of his day, did much to establish the discipline of comparative anatomy, comparing living animals with those known only from fossils. His comparison of living elephants with mastodons and mammoths, found as fossils in Siberia and North America, showed that the fossil animals lived in a temperate or cold environment. This contradicted other scientists who interpreted the fossils to mean that the climates of Siberia and North America had once been warm enough to support close relatives of elephants. It did nothing, however, to undermine the theory that fossils provided clues to past climates. Alexandre Brongniart found fossil evidence in the Paris basin that indicated the climate in the early part of the Tertiary had been warmer than that of the 19th century.

Adolphe Brongniart continued this aspect of his father's work. He studied the coal beds of France. These were formed during the Carboniferous period, 359.2 to 299 million years ago, from plant material that had fallen into shallow, muddy water and subsequently been compressed. Coal contains many plant fossils, and Brongniart concluded from his study of them that coal formed under tropical conditions. This implied that wherever coal beds occurred, even in northern Europe, the climate at the time had been tropical. No one then imagined that continents could move across the Earth's surface, and Brongniart took his discovery to mean that the world's climates had been much warmer during the Carboniferous and that they had been cooling since that time. He described his findings and explained his interpretation in his *Histoire des Végétaux Fossiles* as part of a wider account of the history of plant life in which he related extinct plants to the living ones they most closely resembled. This gave a coherent and largely accurate structure to the study of plant fossils, providing a sound base on which later paleobotanists could build. In addition to his paleobotanical work, Brongniart studied the structure of the sexual apparatus of plants, fertilization, and plant cells.

Brongniart was born in Paris on January 14, 1801. He studied medicine and qualified as a physician, and, although he never practiced, from 1828 to 1830 he taught medicine at the University of Paris. He became a member of the Académie des Sciences in 1834. In 1824 he helped found *Annales des Sciences Naturelles* (Annals of the natural sciences), a scientific journal, and in 1854 he founded the Société Botanique de France (French botanical society) and was its first president. He was professor of botany at the Muséum d'histoire naturelle (Natural history museum) in Paris from 1833 until his death there on February 18, 1876.

CHARLES DARWIN AND EVOLUTION BY MEANS OF NATURAL SELECTION

At his home, Down House, in Kent, Charles Darwin (1809–82) had a large garden that he loved and greenhouses in which he performed experiments. Although his evolutionary theory is most often discussed in its relation to animals, Darwin was every bit as interested in the evolution of plants. He used the way breeders modify animals by selectively breeding those individuals that possess the characteristics the breeders wish to emphasize as a metaphor for evolution by natural selection, and he applied exactly the same argument to plants. In the first chapter of *On the Origin of Species* Darwin wrote the following:

> In plants the same gradual process of improvement, through the occasional preservation of the best individuals, whether or not sufficiently distinct to be ranked at their first appearance as distinct varieties, and whether or not two or more species or races have become blended together by crossing, may plainly be recognised in the increased size and beauty which we now see in the varieties of the heartsease, rose, pelargonium, dahlia, and other plants, when compared with the older varieties or with their parent-stocks. No one would ever expect to get a first-rate heartsease or dahlia from

the seed of a wild plant. . . . I have seen great surprise expressed in horticultural works at the wonderful skill of gardeners, in having produced such splendid results from such poor materials; but the art, I cannot doubt, has been simple, and, as far as the final result is concerned, has been followed almost unconsciously. It has consisted in always cultivating the best known variety, sowing its seeds, and, when a slightly better variety has chanced to appear, selecting it, and so onwards.

Darwin did not discover that the plants and animals living today have evolved from earlier forms. By the middle of the 19th century many biologists had reached this conclusion, but none had proposed a credible mechanism by which evolution might occur. That is the contribution Darwin made. His proposal was simple, persuasive, and although there were many difficulties with it, which he was the first to acknowledge, the vast amount of evidence that has accumulated since his death has resolved those difficulties and demonstrated that his idea was essentially correct. He outlined his theory in *On the Origin of Species by Means of Natural Selection: Or the Preservation of Favoured Races in the Struggle for Life,* published in 1859. The theory can be summarized as:

- There is variation among the individuals of every species.
- Those variations are heritable.
- On average, parents produce more offspring than are needed to replace them (i.e., more than two offspring for each set of parents).
- Populations cannot increase indefinitely and in most populations numbers remain fairly constant.
- It follows that there must be competition among offspring in each generation.
- In that competition, the individuals best fitted to the conditions under which they live will gain better access to resources and tend to produce more offspring.
- Over time, environmental conditions change, and as they do so the qualities that best suit organisms to their environment also change; thus the environmental conditions determine which individuals will produce most offspring. This is natural selection.

🐾 Offspring will inherit the characteristics that equipped their parents for the environments in which they lived.

🐾 The resulting changes will accumulate over many generations, as changing environmental conditions select individuals possessing characteristics that best equip them for the new conditions, until they result in the emergence of new species. This is the origin of species by means of natural selection.

Charles Robert Darwin was born into a prosperous family in Shrewsbury, Shropshire, England, on February 12, 1809. He was the fifth of six children and his parents' second son. His education began in September 1818 when he enrolled as a boarder at Shrewsbury School, joining his brother Erasmus Alvey Darwin (1804–91). In October 1825 Darwin spent the summer helping his father who was a physician, before he and Erasmus both went to the University of Edinburgh to study medicine. Charles learned *taxidermy*—the craft of posing dead animals in lifelike attitudes for display—but found surgery distressing and spent much of his time studying natural history. He left Edinburgh, and in January 1828 entered Christ's College, Cambridge, to commence a course of studies intended to prepare him for becoming an Anglican clergyman. While there he grew friendly with a number of naturalists, including the professor of botany John Stevens Henslow (1796–1861). Darwin did well in his final examination in 1831, and later that year he was accepted as a companion to Robert FitzRoy (1805–65), who was preparing to sail around the world on a surveying expedition as captain of HMS *Beagle*. The voyage lasted almost five years, returning to England on October 2, 1836.

Following his return, Darwin lived for a time in Cambridge and then moved to London. He was under tremendous pressure of work. He had to write and correct his *Journal of Researches into the Natural History and Geology of the Countries Visited During the Voyage of H.M.S.* Beagle *Round*

A photograph of Charles Darwin (1809–82) toward the end of his life *(New York Public Library/Art Resource)*

the World, Under the Command of Captain FitzRoy, R.N., more often known as *The Voyage of the* Beagle, for inclusion in the official report of the voyage. He began editing and preparing the reports on his zoological collections written by various experts as well as Darwin himself. These were published between February 1838 and October 1843 in five volumes as *Zoology of the Voyage of HMS* Beagle *Under the Command of Captain FitzRoy, R.N., during the Years 1832 to 1836.* He had agreed to write a book on geology, and in March 1838, after resisting for a time, he accepted the post of secretary to the Geological Society of London. His health began to deteriorate, and he never recovered. From that time he was prone to attacks of stomach pains and vomiting and suffered from headaches, palpitations, boils, and other symptoms. The cause of his illness was never diagnosed, but the symptoms were especially severe whenever he had to attend meetings or social events, which he found stressful.

On January 24, 1839, Darwin was elected a fellow of the Royal Society, and on January 29 he married his cousin, Emma Wedgwood (1808–96). The delighted family gave them a dowry and investments that brought them in an income on which they could live in comfort for the rest of their lives. In 1842 the couple moved to Down House, bought for them by Darwin's father, and that is where they spent the rest of their lives. Darwin continued with his research and writing. He published *The Descent of Man, and Selection in Relation to Sex* in 1871, *The Expression of Emotions in Man and Animals* in 1872, as well as a number of books on botany, and an autobiography that was published in 1887. The illustration on previous page shows him toward the end of his life. Darwin died at Down House on April 19, 1882. His funeral was held in Westminster Abbey, where Darwin is buried.

ASA GRAY AND THE DISCONTINUOUS DISTRIBUTION OF PLANTS

In 1851 during one of his trips to Europe, Asa Gray (1810–88) visited Kew Gardens, where Joseph Dalton Hooker (1817–1911) introduced him to Charles Darwin. Hooker was director at Kew, and since 1842 Gray had been professor of natural history at Harvard University, where he devoted himself wholly to botany. He was the leading American botanist of the 19th century and an authority on plant distribution. The meeting must have gone well because from 1855 Gray and Darwin conducted a long correspondence, which began with a request from Darwin for information on the distribution of certain North American alpine plants. Gray supplied this, and Darwin made use of his knowledge of plant geography. In 1881 Darwin wrote to Gray that "there is hardly any one in the world whose approbation I value more highly than I do yours." When, on June 18, 1858, Darwin received from Alfred Russel Wallace (1823–1913), another of his many correspondents, a paper entitled "On the Tendency of Varieties to Depart Indefinitely From the Original Type," he realized that Wallace had developed an evolutionary theory very similar to his own, and therefore he had to publish his own theory. After consulting among his friends, Darwin decided that, with Wallace's agreement, he would prepare a statement of his own theory that would be read at a meeting of the Linnean Society together with Wallace's paper. Darwin compiled his paper partly from a letter written in 1857 in which he had outlined his theory to Asa Gray. The papers were duly presented at the meeting, but Darwin was too ill to attend and Wallace was in Asia, so Darwin's friends, Sir Charles Lyell (1797–1875) and Joseph Hooker, read them.

Asa Gray (1810–88), professor of natural history at Harvard University, was Charles Darwin's most influential American supporter. Gray was also the leading U.S. authority on plant classification and plant distribution. *(Science, Industry & Business Library/New York Public Library/Science Photo Library)*

Gray did not agree with Darwin in every respect, but he was Darwin's principal champion in North America, and it was he who arranged for the U.S. publication of *On the Origin of Species* and negotiated the terms with the publisher, D. Appleton & Co. of New York. In 1888 Appleton published Gray's book *Darwiniana: Essays and Reviews Pertaining to Darwinism*, which proved very influential. Later, Gray, who was deeply religious, attempted to persuade Darwin to return to his Christian faith. Darwin appreciated the sentiment but was unable to accede. The two scientists met again in 1868, during a year's leave of absence Gray had taken from Harvard. The photograph at right shows him at about this time.

Asa Gray was primarily a plant taxonomist, but he also engaged in the other major botanical controversy of the 19th century: how to explain

disjunct distribution—the occurrence of closely related species in scattered locations separated by substantial barriers to migration such as oceans. Many scientists believed that land bridges had once extended far from the shores of the continents, allowing plants to extend their ranges into regions that became isolated when the bridges disappeared. Gray had a different idea. He had identified plants growing in Japan that were closely related to plants found in eastern North America, but nowhere in between. He coined the term *vicariance* to describe the occurrence of two closely related species at widely separated locations, so that each species is the geographic equivalent of the other, and he did not believe that vicariance could arise through migration. Plants could not have crossed the Pacific Ocean and North American continent without leaving any traces of their journey. He suggested instead that these Japanese and eastern North American species had originated in the center of North America, from where they had extended their range in both directions until they became widespread. He produced a geological argument to show that conditions might have permitted this on at least two occasions in the past. Ice sheets had then advanced southward across North America, destroying all the plants in their path, but the ice did not extend to the eastern part of the continent, and the plants there survived. When the ice retreated, plants migrated northward to colonize the exposed ground, but the eastern plants remained unchanged. His idea influenced Darwin in the development of his own evolutionary theory.

Gray was born on November 18, 1810, at Sauquoit, New York. He studied medicine at Fairfield Medical School in Connecticut, qualifying in 1831, but he practiced medicine for only a few months. While a medical student, he taught himself botany in his spare time. In 1832 he began teaching science at Bartlett's High School, in Utica, New York, and in 1834 he became an assistant to John Torrey (1796–1873), professor of chemistry and botany at the College of Physicians and Surgeons in New York. Torrey was also a botanist, and he and Gray became lifelong friends and collaborators. Together they compiled the *Flora of North America,* published in two volumes in 1838 and 1843, a work that established them internationally as America's leading botanists. In 1835 Gray was appointed curator and librarian at the New York Lyceum of Natural History, and in 1838 he accepted the professorship of botany at the newly established University of

Michigan. In that year he sailed to Europe to purchase books for the university and to study type specimens of North American plants in European herbaria. Gray spent a year in Europe, and then the opening of the university was delayed so he failed to take up the appointment. In 1842 he accepted the post of professor of natural history at Harvard University on condition that he could devote himself to botany. Gray remained in that position until he retired in 1873.

Gray was one of the first members of the National Academy of Sciences. He became president of the American Academy of Arts and Sciences and of the American Association for the Advancement of Science. He was a regent of the Smithsonian Institution. He wrote more than 360 books, monographs, and papers, one of the most famous being his *Manual of the Botany of the Northern United States, from New England to Wisconsin and South to Ohio and Pennsylvania Inclusive,* which was published in 1848. It is usually called *Gray's Manual* for short. Gray died at his home in Cambridge, Massachusetts, on January 30, 1888.

GÖTE TURESSON AND PLANT ECOTYPES

Natural selection acts on differences that exist among the members of a species, and the concept of such variation is central to Darwin's evolutionary theory. The fact of variation among individuals creates classification problems, however, as taxonomists must decide whether the visible differences between two plants or animals are sufficient to justify classifying them as distinct species. The scientist who first proposed a solution to the dilemma was the Swedish evolutionary botanist Göte Turesson (1892–1970).

Different populations of any widespread species occur in a range of environments that differ slightly in such factors as climate, soil type, and the other plants and animals living there. Species adapt to these local differences in ways that may alter their behavior and sometimes their appearance. For instance, many plants growing in high latitudes flower later in the year than others of the same species growing closer to the equator. When low-latitude plants are moved to a high-latitude environment, they grow less well than they did in their former environment and also less well than other plants of the same species that are already established there. This disadvantage continues into later generations produced from seeds from the transplanted plants, demonstrating that the adaptation has led to a genetic divergence between the two populations. Individuals from one population are still able to interbreed with those from the other population, so they are not different species, but there is a genetic difference between them. Turesson introduced the term *ecotype* to describe a population of a widespread species that is adapted to local conditions in this way.

Turesson's definition suggested that each locally adapted population was in some degree distinct from all other populations of the same species, as though a line could be drawn around it to isolate it. Reality is often different, however. Rather than comprising a patchwork of separate ecotypes, species in most regions shade from one population to the next, with adaptive changes that alter gradually, and the ecotypes occupying the extreme ends. A gradual change of this kind is called a *cline*.

Göte Wilhelm Turesson was born at Malmö, Sweden, on April 6, 1892. He attended school locally and then studied science in the United States at the University of Washington, where he obtained his first degree in 1914 and his master's degree in 1915. Turesson returned to Sweden in 1916 and commenced his research into plant growth. He began teaching at the University of Lund in 1921 and received his Ph.D. in 1922, the year in which he published two essays in the journal *Hereditas*: "The Species and the Variety as Ecological Units" and "The Genotypical Response of the Plant Species to the Habitat." In 1925 he published a third essay, also in *Hereditas*, "The Plant Species in Relation to Habitat and Climate." Turesson was a lecturer at the University of Lund from 1922 to 1927, and from 1927 to 1931 he conducted research at the Weilbullsholm plant breeding center. He returned to teach and research at Lund in 1931, remaining there until 1935, when he failed to obtain a professorship at Lund and accepted the position of professor of botany and genetics at the agricultural college at Uppsala University. He remained at Uppsala until he retired in 1959. Turesson died at Uppsala on December 30, 1970.

NIKOLAI VAVILOV AND THE ORIGIN OF CULTIVATED PLANTS

When early farmers first began to select the plants that they wished to grow, they chose individuals with certain characteristics. In the case of cereals, for example, they selected plants with a tough rachis and a large store of endosperm (see "The Story of Wheat" on pages 130–131). In selecting for certain traits, farmers gradually altered the genetic constitution of their crop plants, and later plant breeders carried the process much further, so today most domesticated plants belong to different species from their wild ancestors. This was the process Darwin used as a metaphor for natural selection.

In the early 20th century a Russian botanical geneticist realized that artificial selection and the geographic spread of agriculture using a limited number of crop species had an interesting implication. Farmers and then plant breeders selected the traits that they considered

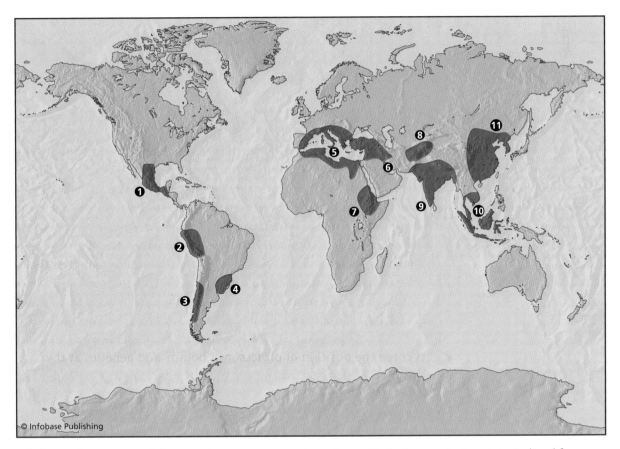

© Infobase Publishing

Nikolai Vavilov proposed these regions as centers where domesticated crop plants originated and from where they had spread. The regions are as follows: 1. Mexico and Guatemala; 2. Peru, Ecuador, Bolivia; 3. southern Chile; 4. southern Brazil; 5. Mediterranean; 6. Middle East; 7. Ethiopia; 8. central Asia; 9. India and Myanmar (Burma); 10. Thailand, Malaysia, Java (Indonesia); 11. China

useful from the variety of traits present in the wild plants. Each time they did so, however, they also rejected all those plants that lacked the desired trait, and they repeated this for trait after trait and generation after generation. Consequently, the amount of natural variation in the crop species slowly diminished. This might pass unnoticed, but agriculture also spread geographically, which suggested that the amount of natural variation in a cultivated species diminished with increasing distance from the region where it was first domesticated. Nikolai I. Vavilov (1887–1943) believed he could measure natural variation and use it to identify the regions where particular crops had first been cultivated. He called those regions centers of origin. They were also centers of diversity, because within them there is the greatest genetic variation in the species in question.

Between 1916 and 1935 Vavilov conducted more than 100 missions to collect plant specimens that he studied in his laboratory at the University of Saratov. He visited Iran, the United States, Central and South America, Afghanistan, the Mediterranean, Ethiopia, China, and South Asia. In 1926 he published *The Centers of Origin of Cultivated Plants.* The following illustration shows that he identified 11 of these. The table that follows lists the crops that originated in each center; some crops were domesticated more than once.

Nikolai Ivanovich Vavilov was born in Moscow on November 25, 1887, into a prosperous merchant family. He studied at the Moscow Agricultural Institute, graduating in 1910, and spent 1911 and 1912 working at the Bureau for Applied Botany and the Bureau of Mycology and Phytopathology. (*Mycology* is the scientific study of fungi; *phytopathology* is the study of plant diseases.) In 1913 and 1914 he continued his studies in England, first at the School of Agriculture at Cambridge University and later at the John Innes Horticultural Institution. After returning to Russia he began to investigate the origin of cultivated plants.

In 1917 Vavilov was made a professor at the University of Saratov. He stayed there until 1921, when he was assigned to the Bureau of Applied Botany in Petrograd (now St. Petersburg). He was director of the Institute of Applied Botany from 1924 to 1929, a full member of the USSR Academy of Sciences, director of the All-Union Institute of Plant Breeding from 1930 to 1940 and of the Institute of Genetics from 1933 to 1940, president in 1929 and vice president from 1935 to 1937 of the Lenin All-Union Academy of Agricultural Sciences, and

CENTERS OF ORIGIN	
CENTER	**CROP**
1. Central America	Maize, common bean, lima bean, jack bean, tepary bean, grain amaranth, Malabar gourd, winter pumpkin, chayote, upland cotton, bourbon cotton, sisal, sweet potato, arrowroot, pepper (capsicum), papaya, guava, cashew, wild black cherry, cherry tomato, cocoa
2. South America	Potato, nasturtium, maize, lima bean, common bean, canna, pepino, tomato, ground cherry, pumpkin, pepper, cotton, cocoa, passion flower, guava, heilborn, quinine, tobacco
3. Southern Chile	Potato, strawberry
4. Southern Brazil	Manioc, peanut, rubber tree, pineapple, Brazil nut, cashew, purple granadilla
5. Mediterranean	Durum wheat, emmer, spelt, oats, clover, flax, rape (canola), black mustard, olive, beet, cabbage, turnip, lettuce, asparagus, celery, chicory, parsnip, rhubarb, caraway, anise, thyme, peppermint, sage, hop
6. Middle East	Einkorn wheat, durum wheat, bread wheat, two-row barley, rye, oats, lentil, lupine, alfalfa, clover, fenugreek, vetch, fig, pomegranate, apple, pear, quince, cherry, hawthorn
7. Ethiopia	Wheat, emmer, barley, sorghum, millet, cowpea, flax, teff, sesame, castor bean, garden cress, coffee, okra, myrrh, indigo
8. Central Asia	Bread wheat, peas, lentil, horse bean, chickpea, mung bean, mustard, flax, sesame, hemp, cotton, onion, garlic, spinach, carrot, pistachio, pear, almond, grape, apple
9. India and Myanmar	Rice, chickpea, pigeon pea, urd bean, mung bean, rice bean, cowpea, aubergine, cucumber, radish, taro, yam, mango, orange, citron, tamarind, sugar cane, coconut, sesame, safflower, tree cotton, jute, crotalaria, kenaf, hemp, black pepper, gum Arabic, sandalwood, indigo, cinnamon, croton, bamboo
10. Thailand, Malaysia, Java	Job's tears, velvet bean, pummelo, banana, breadfruit, mangosteen, candlenut, coconut, sugar cane, clove, nutmeg, black pepper, manila hemp
11. China	Millet, buckwheat, hull-less barley, soybean, velvet bean, Chinese yam, radish, pak soi, onion, cucumber, pear, Chinese apple, peach, apricot, cherry, walnut, litchi, sugar cane, opium poppy, ginseng, hemp

from 1931 to 1940 he was president of the All-Union Geographical Society. He was invited to be president of the International Congress of Genetics held in 1939, and in 1940 he was elected a foreign member of the Royal Society.

In the 1920s, Vavilov befriended Trofim Denisovich Lysenko (1898–1976), a talented young agronomist, and began taking him to scientific meetings. Lysenko was very popular with farmers, and when he declared that he had discovered techniques for hybridizing plants to produce permanent physiological improvements and for fertilizing the fields without using mineral fertilizers, his popularity grew. The USSR was desperately short of food, and the idea that improving the environment could result in heritable improvements in plants fitted the official ideology—that people are the product of their environment—and captured the attention of senior politicians, eventually including Stalin. Lysenkoism also rejected the entire concept of genetics. Lysenko was never able to produce any evidence that his theories worked, but nevertheless he became increasingly powerful and his supporters began to attack the geneticists for generating scientific papers but doing nothing to feed the people. A Congress on Genetics and Agriculture was held in 1936 with the purpose of discrediting the geneticists, and in 1939 a Congress on Genetics and Selection openly attacked Vavilov, who had grown increasingly outspoken in his condemnation of Lysenko's perversion of science. In 1940 Vavilov was arrested and subsequently tried, essentially for opposing the official doctrine. He was sentenced to death, but in 1942 the sentence was commuted to a 20-year prison term. Vavilov's closest colleagues were dismissed from their positions, and many were imprisoned. Vavilov was imprisoned at Saratov, where he died from malnutrition on January 26, 1943.

During the 1950s, following Stalin's death, mainstream scientists began to be more outspoken in their criticisms of Lysenko. This culminated in 1964 with Lysenko's dismissal from his position as director of the Institute of Genetics at the USSR Academy of Sciences. He continued to work at an experimental farm outside Moscow, and the institute was later dissolved. An expert commission was appointed to investigate the records Lysenko kept at the experimental farm. Its report was strongly critical of Lysenko, and it was made public. Lysenko was disgraced, and in the years that followed Vavilov was officially rehabilitated.

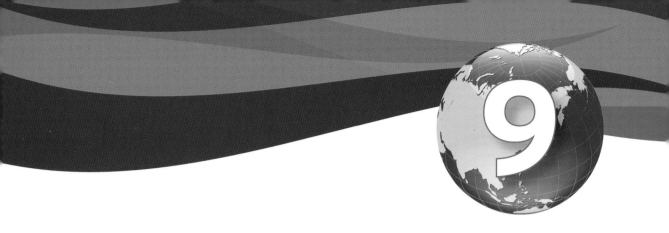

Plant Physiology

P *hysiology* is the scientific study of the way plants and animals work. Plant physiologists concern themselves with the ways plants obtain nutrients and energy and how they reproduce. This branch of botany is closely linked to the study of *anatomy*—the study of the structure of living organisms. While other scientists were devising ways to classify plants, seeking to account for their geographic distribution and unraveling their evolutionary history, the physiologists and anatomists were revealing the innermost details of the plants themselves. This chapter outlines the development of these studies.

The chapter begins with the scientists, one English and the other Italian, whose investigations marked the beginning of the modern study of plant anatomy and physiology. It goes on to tell of the discovery of plant cells, of the way water and nutrients move through plants, and the discovery of oxygen, which was the first step toward explaining respiration. The chapter ends with the introduction of modern cell theory and the internal structure of plant cells.

NEHEMIAH GREW, PLANT REPRODUCTION, AND COMPARATIVE ANATOMY

Scientists who aim to reconstruct past climates and environments make use of *pollen*—the mass of grains produced within the anther of a flower that carry the sperm. Every pollen grain has a tough protective coat that can survive in the soil for many years, and scientists are able to retrieve stored pollen grains from the soil and identify the plants that produced them. That allows them to list the plants that grew in that place at a particular time in the past. Identifying the plants allows the investigators to construct a picture of the environment at that time, and since every plant has its own climatic requirements it also reveals the type of climate that existed then. This work is possible because the size, shape, and surface markings on pollen grains are characteristic of a plant family and often of a genus or even a species. The study of pollen grains and the spores of nonflowering plants is called *palynology,* and it has developed from the work of an English 17th-century plant anatomist and physiologist Nehemiah Grew (1641–1712).

Grew published his most important work in 1682. It was entitled *Anatomy of Plants* and consisted of four books: *Anatomy of Vegetables Begun, Anatomy of Roots, Anatomy of Trunks,* and *Anatomy of Leaves, Flowers, Fruits and Seeds.* It had 82 plates and seven appendixes dealing mainly with botanical chemistry. The work was largely a collection of papers and articles Grew had written previously. Grew maintained that every organ of a plant had two parts, a *ligneous*—woody—part and a pithy part, or parts similar to these. Grew's pithy part was composed of unspecialized plant cells with air spaces between them, a type of tissue for which he invented the name that is still used today, *parenchyma.* Grew also called the rudimentary root that emerges from an embryo the *radicle* and the shoot the plume. The term *radicle* is still used, but the plume is now called the *plumule.*

Chlorophyll is the green pigment that plants use to capture light photons in the first stage of photosynthesis. Nehemiah Grew was the first person to extract chlorophyll from plant tissue; he dissolved it in oil. He noted that the protective scales on buds overlap like the tiles on a roof, and that this arrangement economizes on space. He studied tulip flowers while they were inside their buds, examined the way cotyledons (seed leaves) are folded inside their seed, and found that "bee bread" is made from pollen grains.

Nehemiah Grew was born in September 1641 at Mancetter, a village in Warwickshire, not far from Coventry in the English Midlands. His father, Obadiah Grew (1607–88), was vicar of St. Michael's Church in Coventry. Nehemiah Grew studied at Pembroke College,

University of Cambridge, graduating in 1661, then moved to the Netherlands to study medicine at the University of Leiden, where he qualified as a physician in 1671. He returned to England and practiced medicine in Coventry for a short time, but from 1664 he had been redirecting his training in animal anatomy toward plants. In 1670 Grew wrote *The Anatomy of Vegetables Begun* (later part of his larger work), an essay that caught the attention of Bishop John Wilkins (1614–72), who was the first secretary of the Royal Society. On Wilkins's recommendation Grew was elected a fellow of the Royal Society in 1671. The following year he moved from Coventry to London, where he established a large and successful medical practice. Grew became secretary of the Royal Society in 1677 and was editor of *Philosophical Transactions* in 1678 and 1679. He died in London on March 25, 1712.

MARCELLO MALPIGHI AND THE MICROSCOPIC STUDY OF PLANTS

Nehemiah Grew moved from Coventry to London partly to gain access to the microscopes owned by the Royal Society. Scientists recognized the value of these instruments, and Grew made extensive use of them. It was the Italian physician Marcello Malpighi (1628–94), however, who really pioneered the use of the microscope in the study of anatomy. Malpighi became best known for his microscopic studies of animals, but he also studied plants. In 1671 he published a two-volume work *Anatomia plantarum* (Plant anatomy). The Royal Society published the two volumes in London in 1675 and 1679.

Malpighi had first been intrigued by what looked like fine threads emerging from the branch of a chestnut tree where the branch had been broken. He had earlier discovered the fine tubes, called trachaea, passing through the external skeleton through which air enters the body of an insect, and he assumed these served a similar function in plants. He found that the plant structures were long tubes, thickened at intervals. His observation was accurate, but his interpretation was mistaken. In fact he was looking at xylem—the vessels, made from elongated cells joined end to end, along which water is transported from the roots to every part of the plant.

His enthusiasm aroused, Malpighi watched and drew the stages in the germination of seeds. He was the first scientist to describe the nodules on the roots of leguminous plants, though he was unaware that these were colonies of bacteria that converted atmospheric nitrogen into compounds the plant absorbed in return for carbohydrates the bacteria absorbed from the plant. He studied plant galls and found that some of them contained an insect larva.

Marcello Malpighi was born on his family's farm at Crevalcore, not far from Bologna, Italy, on March 10, 1628. When he was 17, he enrolled at the University of Bologna to study philosophy, but had to interrupt his education for more than two years following the deaths of both his parents and his grandmother. He returned to the university and in 1649 began to study medicine. He graduated in philosophy, qualified as a physician in 1653, and applied unsuccessfully for a post as a lecturer. His application succeeded in 1655, but after a few months he became professor of theoretical medicine at the University of Pisa. Illness forced Malpighi to leave Pisa in 1660, and he returned to Bologna, but left again in 1662 to become the first professor of medicine at the University of Messina for a four-year period, after which he went back to Bologna as professor of medicine, where this time he remained for 25 years. In 1691 he moved to Rome to become private physician to Pope Innocent XII. He died in Rome from apoplexy on November 30, 1694.

ROBERT HOOKE AND THE CELL

In 1665 the English physicist, instrument maker, and inventor Robert Hooke (1635–1703) published a book called *Micrographia* describing his researches using a microscope and illustrated by his own excellent and detailed drawings. Hooke's microscope has survived and is shown in the following illustration. It is now housed at the National Museum of Health and Medicine in Washington, D.C. To adjust the coarse focus, the microscopist could move its tube, covered in leather with gilt tooling, up and down the vertical support, and the objective lens at the bottom could be moved on a screw thread to provide fine focus.

In one of his microscopic investigations, Hooke examined a thin slice of cork and found that it consisted of tiny open spaces bordered by tissue. The spaces reminded Hooke of the monks' cells in a monastery, so he called them cells. That is the origin of the term, but it

The microscope that once belonged to Robert Hooke (1635–1703)

was only one of Hooke's many observations—on optics as well as biological structures—and scholars consider *Micrographia* to be the first important work on microscopy.

Robert Hooke possessed one of the most inventive minds England has ever produced. He was born on July 18, 1635, at Freshwater, on the Isle of Wight, an island close to the southern coast of England. His health was poor, and he spent much of his childhood alone, amusing himself by making mechanical toys. Following the death of his father when he was 13, Hooke began an apprenticeship to a portrait artist, but soon abandoned this and enrolled at Westminster School, and in 1653 he secured a place at Christ Church College, University of Oxford, where he earned money by designing scientific instruments and improving on existing instruments, selling his work to professional instrument makers. While at Oxford, Hooke met many of the leading scientists of the day. In 1660 Hooke was among the group of scientists who formed a society that two years later became the Royal Society. Hooke was employed as curator of experiments at the Royal Society, and in 1663 he was elected a fellow and the same year was made a lecturer in mechanics at the society. In 1665 he also became professor of geometry at Gresham College. Hooke was secretary to the Royal Society from 1677 to 1683. As well as being an inventor, instrument maker, and microscopist, Hooke was also London's official surveyor following the Great Fire of 1666, and he was a successful architect. Hooke never married. He died at his home in Gresham College, London, on March 3, 1703.

STEPHEN HALES, THE MOVEMENT OF SAP, AND TRANSPIRATION

On the surfaces of leaves there are small pores that open and close in response to the movements of two guard cells, one on either side of each pore. The pores are called *stomata* (singular stoma), and they are the openings through which the plant exchanges gases. Carbon dioxide enters the plant and is used in photosynthesis, and oxygen, a by-product of photosynthesis, leaves. While the stomata are open, water in the tissue immediately beneath the leaf surface is lost by evaporation, its place being taken by water drawn upward through the plant. This process is called *transpiration*. Marcello Malpighi was the first microscopist to see and describe the stomata, but he had no idea of their function. The first description of transpiration appeared in 1727, in a book entitled *Vegetable Staticks* written by an English clergyman Stephen Hales (1677–1761). *Vegetable Staticks* was the first volume of Hales's *Statical Essays*. The second volume, called *Haemostaticks*, dealt with animal physiology.

The following diagram shows the process of transpiration and the involvement of the stomata. The evaporation of moisture through the stomata draws up more water to replace it along an unbroken stream that extends all the way through the xylem vessels to the roots. In one of his experiments, Hales cut off a vine at ground level in spring before the buds had opened and attached a long glass tube, mounted vertically, to the severed stem. He found that the sap rose to a height of 24.9 feet (7.6 m). This demonstrated that there was a considerable pressure drawing it upward. His careful measurements of the flow of sap showed that it flows in one direction only, rather than circulating like the blood of a mammal.

This was only one of Hales's experiments with plants. He discovered that the amount of pressure drawing water into the roots varied through the day and according to the temperature. In March 1718 he was elected a fellow of the Royal Society, and in 1719 he presented a paper to the society "Upon the Effect of the Sun's warmth in raising the Sap in trees." He suggested that plants very probably absorb some of their nourishment from the air, and he noted that leaves absorb light, which he found was necessary to their growth.

In addition to his physiological studies, Hales studied air and invented a device that allowed him to breathe only the air he had exhaled. He found he could do this for about one minute before feeling ill, but that if he added an alkaline substance to the apparatus he could continue breathing for more than eight minutes (because the alkali absorbed carbon dioxide, although Hales did not know this). These experiments alerted him to the danger of breathing stale air, and he invented a ventilator to remove stale air—he called it noxious air—from mines, ships, hospitals, and prisons. He considered his ventilator to be his greatest contribution to humanity. He also studied ways to preserve food and water on long ocean voyages and in hot climates.

Stephen Hales was born on September 17, 1677, at Beakesbourne, Kent. In 1697 he entered Bene't (now Corpus Christi) College, University of Cambridge, to study theology, graduating in 1700, receiving

Transpiration is the process by which the evaporation of moisture through leaf stomata generates a pressure that draws up moisture from the soil, through the roots and the plant's xylem vessels.

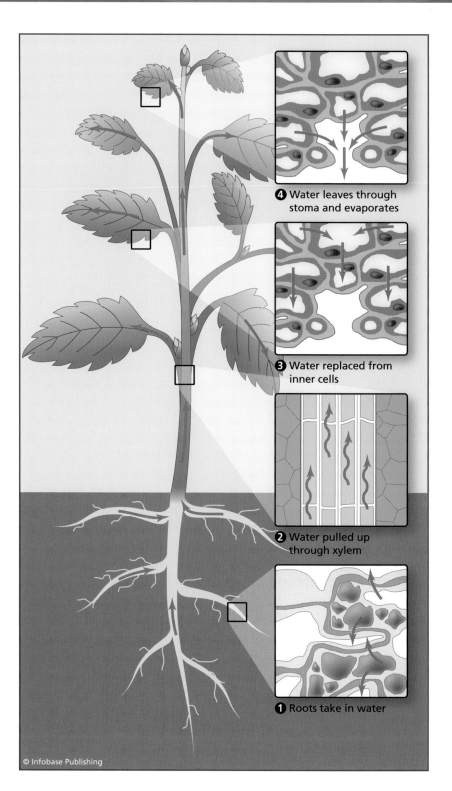

❹ Water leaves through stoma and evaporates

❸ Water replaced from inner cells

❷ Water pulled up through xylem

❶ Roots take in water

his master's degree in 1703, and graduating as a bachelor of divinity in 1711. He received his doctorate of divinity from Cambridge in 1733. While still a student, Hales developed an interest in astronomy, physics, and chemistry, but his attention was directed principally toward the physiology of plants and animals. Hales was ordained in 1703, and in 1709 he left Cambridge—where he had spent 13 years—to take up the position of perpetual curate at Teddington, Middlesex, where he remained for the rest of his life. From 1751 he also held court appointments, first as almoner (the official who distributes charity) to Augusta of Saxe-Gotha, the princess dowager of Wales, and subsequently as chaplain to the princess and to her son, who became George III. Hales was also rector of Porlock, Somerset, and Faringdon, Hampshire, where he spent his summer while spending the rest of the year at Teddington. In 1719 he married Mary Newce, who died in 1721. They had no children. Hales died at Teddington on January 4, 1761.

JOSEPH PRIESTLEY AND "DEPHLOGISTICATED AIR"

In 1772 William Petty, the second earl of Shelburne (1737–1805), invited Joseph Priestley (1733–1804), with his wife Mary and their three children, to live on his estate near Calne, Wiltshire, where Priestley would work as Petty's librarian and tutor to his children. The Priestleys moved to the Shelburne estate the following year. That is where their third son was born, and it is where Priestley made his most famous discovery—of oxygen. He did not call it oxygen, however. It was Priestley's friend, the French chemist Antoine-Laurent Lavoisier (1743–94) who gave the gas its name, *oxygène,* derived from the Greek words *oxu-* meaning "acid" and *-genes* meaning "born." Lavoisier believed (incorrectly) that all acids contain oxygen.

In 1772 Priestley had shown that a gas given off by plants is necessary to animal life. In 1774 he found a way to prepare the gas—or air as he always called it—by heating mercuric oxide (HgO) or minium, also called red lead (Pb_3O_4). When he placed mice inside a sealed container filled with this air, he found they remained conscious for twice as long as they did when the container held only ordinary air. He tried breathing the air himself and found it superior to ordinary air. By heating these oxides, Priestley had driven off the breathable air (oxygen), but that is not what he believed had happened. Priestley believed in the phlogiston theory (see the following sidebar), according to which roasting the metal (mercury or lead) in air had driven off the phlogiston it contained, leaving behind a substance known as the metallic calx. The heating process was known as calcination. With certain metals, heating the calx allowed phlogiston to enter the calx, thereby phlogisticating it and restoring the original metal. Air containing too much phlogiston was unbreathable, so removing the phlogiston by phlogisticating the mercury or lead calx improved the air quality. Priestley called the product of this process "dephlogisticated air." Lavoisier did not believe in phlogiston—it was one matter on which the friends disagreed—and he was able to recognize the gas as a chemical element and a natural constituent of ordinary air.

Dephlogisticated air, or oxygen, was only one of the gases Priestley discovered. He also isolated nitrous air (nitric oxide, NO), alkaline air (ammonia, NH_3), acid air (hydrochloric acid, HCl), and dephlogisticated (also called diminished) nitrous air (nitrous oxide, N_2O). He later discovered vitriolic acid air (sulfur dioxide, SO_2) and also isolated carbon monoxide (CO), but failed to recognize it as a distinct air.

Priestley also invented soda water, publishing a description of the method he used for the benefit of the crew sailing on James Cook's (1728–1779) second voyage to the South Seas—he mistakenly believed it would cure scurvy. Priestley made no money from soda water, but a German silversmith and watchmaker called Johann Jacob Schweppe (1740–1821) patented the process in 1783.

Joseph Priestley was born on March 13, 1733, in the small town of Birstall, West Yorkshire, about 6 miles (10 km) from Leeds, the oldest of the six children of Jonas Priestley, a finisher of cloth, and his wife, Mary Swift. His mother died when Joseph was six, and when his father remarried in 1741 the boy went to live with his wealthy uncle and aunt, John and Sarah Keighley. He attended local schools where he learned Latin, Greek, and Hebrew. He also studied French, German, Italian, Chaldean, Syrian, and Arabic. The family were Calvinists and therefore religious Dissenters, and his education continued at a Dissenting academy (a school for the children of Dissenters) in Daventry, Warwickshire. He matriculated in 1752, and in 1758 he became a clergyman, with parishes in Needham Market, Suffolk, and later at Nantwich, Cheshire. In 1761 he was appointed to teach modern languages and rhetoric at the Dissenting Warrington Academy

PHLOGISTON

In 1667 the German chemist Johann Joachim Becher (1635–1682) published a book called *Physica Subterranea* (Physics below ground), in which he revised the traditional view of the classical elements. Becher replaced the elements fire and earth with three alternative forms of earth to which he gave Latin names: *terra lapidea,* or "stony earth," which was the quality allowing earth to fuse into a solid mass, *terra fluida,* or "flowing earth," governing the ease with which earth will flow, and *terra pinguis,* or "fatty earth," which is concerned with combustion. Becher argued that when any combustible substance is burned *terra pinguis* is released.

One of Becher's students was another German chemist Georg Ernst Stahl (1660–1734). Stahl expanded Becher's ideas and renamed *terra pinguis,* calling it phlogiston, from the Greek word *phlogos* meaning "flame." Phlogiston was colorless, odorless, tasteless, and could not be sensed by touch, but it possessed mass. Every combustible substance contained it and released it when it was burned, together with caloric (heat). The residue, after burning, was called calx (plural calces). Calx was the true form of the substance.

Prior to burning, a substance was said to be "phlogisticated" and after burning the calx was "dephlogisticated." Calx weighed less than the original phlogisticated substance because it had lost phlogiston. Different substances left behind different amounts of calx depending on the amount of phlogiston they contained. Charcoal and sulfur leave very little calx because they are almost pure phlogiston.

Burning released phlogiston into the air, and when substances were burned in an enclosed space the air became increasingly phlogisticated. A point could be reached where the phlogisticated air was incapable of supporting further combustion because it was saturated with phlogiston. What is more, animals could not survive in fully phlogisticated air, because respiration became impossible. Respiration, therefore, removed phlogiston from the body.

The phlogiston theory was highly successful because for more than a century it provided a plausible explanation for natural phenomena and experimental results. Chemists believed it, and it was not until late in the 18th century that it was finally disproved and quickly abandoned.

in Cheshire. Until the laws were repealed in 1829, Roman Catholics and Dissenters, also known as Nonconformists although the terms are not strictly synonymous, were discriminated against in England. They were not permitted to hold public office, stand for election to Parliament, serve in the army or navy, or attend the universities of Oxford or Cambridge.

On June 23, 1762, Joseph married Mary Wilkinson. While at Warrington, Priestley conducted scientific experiments, mainly on electricity, and lectured on anatomy. It was at this time that he met Benjamin Franklin (1706–90), who encouraged his growing interest in science. In 1767 Joseph, Mary, and their daughter, Sarah, moved to Leeds where Joseph became minister of Mill Hill Chapel. They remained there until 1773, and two sons, Joseph and William, were born during their time there. Their youngest son, Henry, was born at Calne in 1777. While he was at Mill Hill, Priestley sent five scientific papers to the Royal Society, describing his experiments with electricity and optics. The Royal Society awarded him their Copley Medal in 1773.

The family moved to Birmingham in 1780, following a disagreement with Lord Shelburne. Priestley became a minister again, and he also joined the Lunar Society, a group of scientists, engineers, inventors, and manufacturers who met once a month when the Moon was full, to minimize the risk of being attacked on the unlit streets. For many years Priestley had devoted considerable effort to theological and political campaigning, which made him a controversial figure. Their support for the American and French Revolutions had made the Dissenters increasingly unpopular, and in 1791 riots broke out in Birmingham. Joseph and Mary Priestley fled from their home, which was attacked and burned to the ground, destroying all their possessions and Joseph's laboratory. Cartoons were published attacking him, and an effigy of him was burned. He was forced to resign from the Royal Society, was attacked in speeches in Parliament, denounced by preachers, and his sons were unable to find work. The sons decided to emigrate to America, and, although Joseph had been made an honorary citizen of France, he and Mary decided to accompany them, escaping shortly before the government began arresting those who spoke out against them. The Priestleys sailed for America on April 17, 1794, arriving to a warm welcome in New York. They then moved to Philadelphia, where Joseph was offered, but declined, the professorship of chemistry at the University of Pennsylvania. Instead, the family moved to the town of Northumberland, where their son Joseph and others were establishing a colony for English Dissenters. Henry Priestley died in 1795, and Mary died the following year. Joseph's health deteriorated, and by 1801 he was unable to work. He died at Northumberland on February 6, 1804. He had been elected to the membership of every leading scientific society in the world.

ERASMUS DARWIN AND *THE BOTANIC GARDEN*

The Lunar Society boasted several members of outstanding intellect, who quite cheerfully referred to themselves as lunaticks. Joseph Priestley was one, and another was Erasmus, the grandfather of Charles Darwin. Erasmus Darwin (1731–1802) believed that in the natural world species were constantly developing as they struggled to overcome the constraints imposed by their environment. He summarized his view of continual change in the following passage from his book *Zoönomia, or, The Laws of Organic Life,* published between 1794 and 1796.

> Would it be too bold to imagine that, in the great length of time since the earth began to exist ... that all warm-blooded animals have arisen from one living filament, which the great First Cause endued with animality, with the power of acquiring new parts, attended with new propensities, directed by irritations, sensations, volitions and associations, and thus possessing the faculty of continuing to improve by its own inherent activity, and of delivering down these improvements by generation to its posterity, world without end!

This picture of continual change is close to the theory advanced by Lamarck (see "Jean-Baptiste Lamarck and the Royal Garden, Paris" on pages 74–76) that by struggling against environmental constraints organisms modify their bodies and their offspring inherit those modifications. This leads to the emergence of new species, but in a progressive way driven by every organism's desire for self-improvement. Darwin did not develop it into a complete theory of evolution, and his ideas are very different from those of his grandson.

Erasmus Darwin was a physician, scientist, and inventor (he designed a rocket powered by hydrogen and oxygen, although it was never built), but during his lifetime he was best known as a poet. He was born, the youngest of seven children, on December 12, 1731, at Elston Hall, in Elston, Nottinghamshire. He was educated at Chesterfield Grammar School and St. John's College, University of Cambridge, and he studied medicine at Edinburgh Medical School. Having qualified as a physician, in 1756 Darwin settled in Nottingham and tried to establish a medical practice. This was not very successful, so in 1757 he moved to Lichfield, Staffordshire, where his success in saving the life of a patient who had been close to death made him popular and his practice prospered. Darwin married twice and had 14 or possibly 15 children, two or three of them illegitimate. After his second marriage he and his wife moved to Radbourne Hall near Derby and later into Derby itself. He died suddenly on April 18, 1802, soon after moving again, this time to Breadsall Priory, near Derby.

While he was in Lichfield, Darwin and several of his friends, one of whom was Samuel Johnson (1709–84), formed the Botanical Society of Lichfield, with the purpose of translating the works of Linnaeus from Latin into English. The task took seven years and ended with the publication of two books, *A System of Vegetables* in 1783 and 1795, and *The Families of Plants* in 1787. His careful study of Linnaeus inspired Darwin to write his two most famous poems. The first was *The Loves of the Plants*, which Darwin published anonymously in 1789. Its four cantos expounded the Linnaean system of classification, based on the reproductive structures of plants, and in it Darwin described female and male plants as though they were human brides and bridegrooms. A "botanic muse" guides the reader and between each of the cantos there is a dialog between the poet and the bookseller. *The Loves of the Plants* proved instantly successful, and this encouraged Darwin to republish it, but this time adding another and more demanding poem, *The Economy of Vegetation*, in which Darwin celebrated scientific and industrial progress. Despite its title, much of it described mining, steel manufacture, and contemporary scientific theories. The two poems were published together in 1791, with *The Economy of Vegetation* preceding *The Loves of the Plants*, as *The Botanic Garden*. It was an immediate best seller.

The Botanic Garden was unique. No one had attempted to write a botanical text in heroic couplets, and no one has done so since. The poem describes 83 plant species, with extensive footnotes explaining the imagery. Darwin's stanza on the sundew (*Drosera*) gives a flavor of the work. Sundew is a carnivorous plant of peat bogs, where nutrients are scarce. Its leaves bear red-tipped hairs, each with a drop of a sticky substance that looks like dew—hence the name. An insect alighting on the plant to drink the dew is trapped. This is Darwin's description, in Canto I of *The Loves of the Plants*:

> Queen of the marsh imperial Drosera treads
> Rush-fringed banks, and moss-embroider'd beds
> Redundant folds of glossy silk surround

Her slender waist, and trail upon the ground;
Five sister-nymphs collect with graceful ease,
Or spread the floating purple to the breeze;
And *five* fair youths with duteous love comply
With each soft mandate of her moving eye.
As with sweet grace her snowy neck she bows,
A zone of diamonds trembles round her brows;
Bright shines the silver halo, as she turns;
And, as she steps, the living lustre burns.

MATTHIAS SCHLEIDEN, THEODOR SCHWANN, AND CELL THEORY

It was Robert Hooke in the 17th century who first observed cells and gave them that name (see "Robert Hooke and the Cell" on pages 166–168), but the German botanist Matthias Schleiden (1804–81) was the first scientist to appreciate their importance. All living organisms either consist of a single cell or are made up of cells, and organisms grow and reproduce by the division of cells. This fundamental tenet of biology is called the *cell theory*. It was first stated in 1838 in a book by Schleiden entitled *Beiträge zur Phytogenesis* (Contributions of phytogenesis).

Schleiden based his conclusion on observations of plant tissues. Hooke had examined the dead tissues he found in cork, but Schleiden studied living cells and he saw that their contents moved within and between the cells and along fibers composed of elongated cells joined end to end. Schleiden called this process *protoplasmic streaming;* the protoplasm outside the cell nucleus that he saw is now known as *cytoplasm.* Schleiden also described the division of the cell nucleus during cell division, but mistakenly thought a daughter nucleus separated from the parent nucleus by budding. Nevertheless, his work gave biologists their first insight into the most basic structure of all living organisms.

In preparing his theory, Schleiden had consulted his friend the German physiologist Theodor Schwann (1810–82), and the following year, 1839, Schwann extended the cell theory to animals, in *Mikroskopische Untersuchungen über die Übereinstimmung in der Struktur und dem Wachstum der Tiere und Pflanzen* (Microscopical researches into the agreement between the structure and growth of animals and plants). Schleiden and Schwann are jointly credited with having originated the cell theory. Schwann was also the first scientist to observe that an egg begins as a single cell and develops into a complex organism by repeated cell division.

Matthias Jakob Schleiden was born in Hamburg on April 5, 1804. In 1824 he entered the University of Heidelburg to study law. He graduated in 1827, and for a time he practiced law in Hamburg, but then turned to botany and medicine, which he studied at the universities of Göttingen, Berlin, and Jena, finally graduating in 1831. After graduating Schleiden was appointed professor of botany at Jena, where he remained until 1862, when he became professor of botany at the University of Dorpat, Estonia. In 1864 he returned to Germany and began teaching privately in Frankfurt-am-Main, where he died on June 23, 1881.

Theodor Schwann was born on December 7, 1810, at Neuss, not far from Düsseldorf, Germany. He was educated at the Jesuit college in Cologne and studied medicine at the universities of Bonn, Würzburg, and Berlin. He qualified in medicine at Berlin in 1834. After graduating he spent four years working as an assistant to the physiologist Johannes Müller (1801–58) at the Museum of Anatomy in Berlin. In 1836 and 1837 Schwann studied fermentation and was able to show that the fermentation of sugar to alcohol was the result of processes within living yeast cells. This work came in for heavy criticism, and in 1839 Schwann left Germany to become professor of anatomy at the Roman Catholic University of Louvain, Belgium. He remained there until 1848, when he became professor of anatomy at the University of Liège. He died in Cologne on January 11, 1882.

Schwann strongly refuted the idea of *spontaneous generation*—that living animals could emerge from putrefying matter. The cell theory supported this refutation, and the theory was encapsulated in an epigram *omnis cellula e cellula* (every cell from a cell) by the French naturalist and physiologist François-Vincent Raspail (1794–1878). The German physician and biologist Rudolf Virchow (1821–1902) popularized the epigram in 1858, and Virchow is sometimes included as one of the originators of the cell theory.

ROBERT BROWN, THE CELL NUCLEUS, AND THE STUDY OF POLLEN

Charles Darwin's theory of evolution by natural selection was first presented in the form of a paper read at a meeting of the Linnean

Society in London on July 1, 1858. The meeting had been arranged hastily, following Darwin's receipt on June 18 of a paper by Alfred Russel Wallace (1823–1913) setting out an almost identical theory. Meetings of the Linnean Society are arranged months if not years ahead, but Darwin was lucky because there had been a cancellation necessitated by the death on June 10 of the advertised speaker, the Scottish botanist Robert Brown.

At the time of his death Brown was curator of the botanical collections at the British Museum, a large part of which had formerly belonged to Sir Joseph Banks (see "Sir Joseph Banks, Unofficial Director of Kew" on pages 70–72). Brown had been Banks's librarian, and in his will Banks had bequeathed to Brown the full use of his books and specimens for life, but on condition that the collection be stored at the British Museum.

In 1827, the year Banks's material was transferred to the museum, Robert Brown had been studying plant pollen from pinkfairy (*Clarkia pulchella*). A keen microscopist, Brown had suspended the pollen grains in water to make them easier to observe. He found that the grains moved about ceaselessly in an agitated fashion and that when he repeated the observation using fine grains of carbon and metal those moved in the same way, proving that the motion was not due to any biological process. He described his discovery in 1828 in an article "A brief account of microscopical observations made in the months of June, July and August 1827 on the particles contained in the pollen of plants, and on the general existence of active molecules in organic and inorganic bodies," published in the *Edinburgh New Philosophical Journal*. Brown had no idea what caused the movement. Physicists now know that it results from the random movement of molecules, and it is known as Brownian motion.

Brown also studied the anatomy of fossilized plants and plant reproduction. In 1831 while he was investigating plant fertilization, he noticed a small structure that occurred in all plant cells and that played an important part in cell division. He called it a nucleus, and that is the name by which it continues to be known.

Robert Brown was born at Montrose, Scotland, on December 21, 1773, the son of an Episcopalian clergyman. He was educated at Marischal College, Aberdeen, and studied medicine at Edinburgh University, but without obtaining a degree. In 1795 he joined the Fifeshire Regiment of Fencibles as a surgeon's mate with the rank of ensign and almost immediately the regiment was posted to Ireland. Brown was a hard worker, with a strict routine. Before breakfast he studied German, after breakfast he studied botany, he saw patients for two hours during the afternoon, and unless he was socializing he continued his scientific studies until midnight.

In 1798 Brown was sent to London to find recruits, and while there he was introduced to Banks, who was impressed by him. When, in 1800, Banks was planning a voyage of discovery to Australia he chose Brown to accompany him as a botanist. They sailed on July 18, 1801, on the *Investigator*, commanded by Captain Matthew Flinders, and reached Australia on December 8. They arrived back in England in October 1805 with specimens of nearly 4,000 species of plants, as well as many zoological specimens. The government paid Brown a salary during the five years he worked on the material. He described 2,200 species, of which 1,700 were previously unknown to science, and Brown named 140 new genera. From 1806 to 1822 Brown worked for the Linnean Society. He was elected a fellow of the Linnean Society in 1822 and was its president from 1849 to 1853. He died in London.

Ecology of Plants

Plants do not grow in isolation. Even in a semidesert, where large expanses of bare ground separate isolated clumps of plants, there are usually several species growing in those clumps. Plants growing close together exploit the environment in different ways. Some have deeper roots than others, to find water and nutrients at lower levels, while those with shallow roots derive nutrients from organic material decomposing near the surface.

By the second half of the 19th century, botanists had come to recognize that knowledge of the evolutionary history and relationships of plants and understanding of their physiology did not tell the whole story. In 1866 the German zoologist Ernst Heinrich Philipp Haeckel (1834–1919) published a two-volume work entitled *Generelle Morphologie der Organismen* (General morphology of organisms). Haeckel had read Darwin's *On the Origin of Species* and was an enthusiastic supporter of the theory of evolution by natural selection. Indeed, he had written a popular book expounding it, *Natürliche Schöpfungsgeschichte* (The natural history of creation). In *Generelle Morphologie* Haeckel sought to explore the implications of the theory. In doing so he coined a new word, *Ökologie,* to describe the web of relationships between living organisms and their physical, chemical, and biological surroundings. He applied the term principally to animals, but it applied equally to plants. *Ökologie* was transliterated into English as oecology, but following the Botanical Congress held in Madison, Wisconsin, in 1893, a group of botanists who had attended the congress agreed to standardize the spelling as *ecology,* and that is how the word has been spelled ever since. Ecology is the scientific study of the relationships among living organisms and between living organisms and their living and nonliving environment.

This chapter outlines the rise of plant ecology. It describes the way botanists learned to classify plants not by their evolutionary relationships, but according to the way they are adapted to the climatic conditions in which they grow. This led them to develop the concept of plants as communities and to the study of the many different types of plant communities.

CHRISTEN RAUNKIÆR AND THE WAY PLANTS GROW

In 1903 the Danish botanist Christen Raunkiær proposed a solution to the difficult botanical problem of comparing plant communities with entirely different compositions. Raunkiær's idea was to categorize plants by the position of their *perennating buds*—the plant structure with which a plant survives periods of adverse conditions. Raunkiær believed that flowering plants first appeared in the Tropics, and that they then spread from there into higher latitudes. In doing so, plants had to evolve strategies for surviving periods of cold or very dry weather. Different plants developed different strategies, and it was those strategies that he used as the basis of his categorization.

At the onset of the cold or dry season, the parts of a plant that are above ground level die back. Woody plants such as trees and shrubs lose their leaves and appear to be dead. Nonwoody plants wither and appear to die. The plants are not dead, however, and they store food in tubers, bulbs, rhizomes, or in the case of woody plants as buds. These are the perennating buds from which the plant will grow back when the weather improves. A tuber is a swollen stem or root that serves as an underground storage organ. A *bulb* is an underground storage organ consisting of a short, fat stem with roots at its base above which there are fleshy leaves surrounded by protective scales. A *rhizome* is a horizontal, creeping, underground stem from which roots and shoots emerge at intervals. All of these are perennating buds, and Raunkiær identified five types, with some subdivisions, each of which offered a different degree of protection, depending on its distance from ground level. In order of the protection the position of the perennating buds affords, the Raunkiær categories are as follows:

Phanerophytes are plants in which the perennating buds are above ground on shoots exposed to the air. This affords the least protection from harsh weather, and phanerophytes are found where cold or dry conditions occur infrequently. They are trees and shrubs and also *epiphytes*—plants that grow on the surfaces of other plants, using those plants only for support and not as a source of nutrients.

Chamaephytes are plants in which the perennating buds occur very close to the ground. This offers rather more protection.

Hemicryptophytes are plants in which the perennating buds are at ground level.

Cryptophytes are plants in which the perennating buds are below the ground or water surface. There are three types of cryptophytes. *Geophytes* have perennating buds below the ground. *Helophytes* have perennating buds below the surface of a marsh. *Hydrophytes* have perennating buds below the surface of water.

Therophytes are plants that complete their life cycles rapidly when conditions are favorable, then die and survive as seed until favorable conditions return. These are annual plants that are also known as *ephemeral plants* or *ephemerophytes.*

Raunkiær identified his categories by comparison with the world average strategy, which he called the normal spectrum, rather than the strategy found in the Tropics. The system he devised is still widely used. He also discovered that when the plant species growing in a specified area are counted and the numbers of the species are arranged in frequency classes, each comprising 20 percent of the total, plants were either very common or very rare. This came to be known as *Raunkiær's law.* Raunkiær believed that everything could be counted and understood statistically.

Christen Christensen was born on March 29, 1860, at Lynhe in western Jutland, Denmark, on a large farm called Raunkiær. At that time it was customary in Scandinavia for people to have a first name and patronymic, but a person enrolling at a university had to provide a surname, so he took the name of the farm. Raunkiær studied at the University of Copenhagen, graduating in botany in 1885. From 1885 to 1888 he taught at Borchs Botanical College, and in 1893 he obtained a position as a scientific assistant at the Copenhagen

University botanical gardens and museum. In 1911 he succeeded Eugen Warming (see "Eugen Warming and the Principles of Plant Ecology" on pages 189–191) as professor of botany and director of the botanical gardens at Copenhagen, remaining in this post until 1923. He died in Copenhagen on March 11, 1938.

JOSIAS BRAUN-BLANQUET AND THE SOCIOLOGY OF PLANTS

Raunkiær based his system on the way plants have adapted to climate. Other botanists were using a similar approach to classify large units of vegetation, and one of the most influential was the German botanist and plant geographer August Grisebach (see "August Grisebach and Floral Provinces" on pages 122–124). Other European plant ecologists developed Grisebach's ideas. This led to the emergence of phytosociology as the study of plant communities. The foremost phytosociologist was the Swiss ecologist Josias Braun-Blanquet (1884–1980). Braun-Blanquet assembled a team of biologists who came to be known first as the Zurich School of Phytosociology, and from 1930 as the Zurich-Montpellier School of Phytosociology. The change of name reflected the team's move from Zurich, Switzerland, to Montpellier, France, where Braun-Blanquet had been appointed director of the newly established International Station for Alpine and Mediterranean Botany, at Montpellier. The institution's French name is Station Internationale de Géobotanique Méditerranéenne et Alpine. This long name is usually abbreviated to SIGMA, and the Zurich-Montpellier approach to phytosociology—which continues to thrive—is sometimes known as Sigmatism.

In 1928 Braun-Blanquet published a book entitled *Pflanzensoziologie* (Plant sociology), based on his studies of the plant species he had found on Mount Aigoual in the Cévennes region of France. The book proved highly influential, and he revised it for its third edition in 1964. In his book Braun-Blanquet set out a method for classifying plant communities. He and his colleagues began developing and applying this scheme at the University of Zurich in Switzerland, and from 1930 at Montpellier. The method they devised began by surveying an area to identify the most characteristic species among its vegetation. The characteristic species indicated the type of environment, and that species together with the other plants in the area constituted a *relevé*. Relevés were then ranked hierarchically according to their geographical extent, with a phytosociological class occupying the largest area. The ranks in the hierarchy were identified by adding a specified ending to the stem of the name of the characteristic genus.

Josias Braun-Blanquet was born as Josias Braun at Chur, Switzerland, on August 3, 1884. He trained to be a shopkeeper while studying the plants around his home in his free time. The experience he gained observing the local flora allowed him to find employment from 1905 to 1912 as an assistant to the Swiss plant ecologists Heinrich Brockmann-Jerosch, Eduard Rübel, and Carl Schröter. From 1913 to 1915 he studied at the University of Montpellier under the French botanist Charles Flahault (1852–1935). In 1915 Braun received his Ph.D. from the University of Montpellier. The same year he married a fellow student Gabriella Blanquet. Following his graduation, Braun-Blanquet as he then was became an assistant to Rübel until 1922, when he obtained the post of lecturer at the Cantonal Technical High School in Zurich. He remained at Zurich until 1926, during which time he formed the Zurich School. In 1926 he returned to live in Montpellier, earning his living as a private teacher. He was appointed the first director of SIGMA in 1930 and remained in that position until his death at Montpellier on September 20, 1980.

GUSTAF DU RIETZ AND COMMUNITIES OF PLANTS

Braun-Blanquet and his colleagues founded one school of phytosociology, but there were others. Professor Teodor Lippmaa (1892–1943) established an Estonian school in 1934 at the University of Tartu, and its first task was to map the plant distribution throughout the country. This was completed in 1955. The plant communities were then grouped into associations on the basis of soil type, and the associations were grouped according to the predominant plant species and the amount of water available to them, so the grouping proceeded from the most arid to the wettest environments.

The principal rival to the Zurich–Montpellier School was founded in 1921 by Gustaf Du Rietz (1895–1967) at the University of Uppsala, Sweden. The Uppsala School based its studies on observations that Du Rietz had made. He found that some combinations of plant species occurred more frequently than other combinations; he called these associations. There was usually a distinct boundary or narrow transition zone between an association defined by one dominant species and its neighbors. Patches of different associations sometimes occurred within the same habitat. Species that were not dominant occurred in some associations but not in others. In a word, the distribution of species appeared somewhat random. Du Rietz explained this as the result of competition between dominant species in adjacent associations. Competition led to the elimination of one of the species, but which would win depended on their relative abilities at reproducing in the surrounding associations where neither was dominant. This implied that competition between plants could produce stable, long-lasting associations with quite different compositions. The Uppsala ecologists sampled areas of habitat by measuring out areas, called *quadrats,* commonly one square meter in area but larger in woodland, and counting every plant within the quadrat. Quadrats were marked out in random locations.

For a time the Zurich-Montpellier and Uppsala Schools offered alternative phytosociological systems, each with its own technical terms. This was obviously confusing, and the two approaches were eventually reconciled. The two schools still exist, but now they differ only in the emphasis they place on different aspects of their classifications. The Nomenclature Commission of the International Association for Vegetation Science now regulates the vocabulary used in phytosociology.

Gustav Einar Du Rietz was born on April 27, 1895, at Sandvik, Stockholm. He studied at the University of Uppsala, where he received his Ph.D. In 1934 he became professor of plant ecology at Uppsala. He died in Uppsala on March 7, 1967.

ANDREAS SCHIMPER AND PLANT ADAPTATION TO THE ENVIRONMENT

Agriculture and commercial forestry have transformed the landscape over almost the whole of Europe. There are few areas of true wilderness remaining, and in an ecological sense the plant communities are not those that would have developed without human interference. Consequently, it is not easy to observe the way plants have adapted naturally to the climate. The German botanist and phytogeographer Andreas Schimper (1856–1901) pointed this out in 1898 in his book *Pflanzengeographie auf Physiologischer Grundlage,* published in English in

1903 with the title *Plant Geography on a Physiological Basis,* as the following excerpt from the English edition demonstrates.

> The greater prominence of physiology in geographical botany dates from the time when physiologists, who formerly worked in European laboratories only, began to study the vegetation of foreign countries in its native land. Europe, with its temperate climate and its vegetation greatly modified by cultivation, is less calculated to stimulate such observations; in moist tropical forests, in the Sahara, and in the tundras, the close connexion between the character of the vegetation and the conditions of extreme climates is revealed by the most evident adaptations.

It was Schimper, in *Plant Geography on a Physiological Basis,* who first used the term *tropical rain forest* to describe one type of vegetation that is adapted to a warm, humid climate. He defined tropical rain forest as being: "evergreen, hygrophilous in character, at least thirty meters high, rich in thick-stemmed lianes, and in woody as well as herbaceous epiphytes." It is the definition that remains in use today. *Lianes* or *lianas,* are free-hanging climbing plants. *Hygrophilous* means growing in or preferring moist habitats.

Andreas Franz Wilhelm Schimper was born in Strasbourg, France, on May 12, 1856, where his father, Wilhelm Philipp Schimper (1808–80), was director of the natural history museum and a professor of geology. Andreas was educated at the Strasbourg Gymnasium (high school) from 1864 to 1874, when he enrolled at the University of Strasbourg to study biology. He received his Ph.D. at Strasbourg in 1878 and spent a year, 1879–80, at the University of Würzburg, working with the botanist Julius von Sachs (1832–97), who had studied the process by which plants orient themselves toward light. Schimper was a fellow at Johns Hopkins University from 1880 to 1882 and visited the West Indies and Venezuela before returning to Europe in 1883 to take up a post as lecturer at the University of Bonn.

While he was at Bonn, Schimper published the results of his physiological researches. He was the first person to describe *chloroplasts*—the bodies within plant cells that contain chlorophyll and that are the sites of photosynthesis. Schimper called them *Chlorophyllkörner* (chlorophyll grains) and *Chlorophyllkörper* (chlorophyll bodies). He then changed direction, concentrating on phytogeography and plant ecology. In 1886 the University of Bonn made Schimper an extraordinary professor (a professor who does not occupy a chair). He remained at Bonn until 1899, when he became professor of botany at the University of Basel, Switzerland, a position he held until his death in Basel on September 9, 1901.

CARL GEORG OSCAR DRUDE AND PLANT FORMATIONS

Tropical forests extend across a vast area of Central and South America, West and Central Africa, South Asia, and northern Australia. A belt of coniferous forest stretches across northern Canada and Eurasia. The North American prairies, South American pampas, and Eurasian steppe are temperate grasslands. It would be easy to suppose that a visitor might travel through any one of these vast expanses and notice very little change in the vegetation. The forest or grassland would continue for hundreds of miles. In fact, it is not like that. Although the vegetation may all be of one general type such as coniferous forest or temperate grassland, there are many local variations. The trees and grasses in one area are different from those in another.

In 1896 the director of the Dresden Botanical Gardens published a book entitled *Die Ökologie der Pflanzen* (The ecology of plants) in which he showed how local factors such as hills and valleys and local variations in climate influenced the composition of plant communities. That explained why there are local variations. The author of the book was Oscar Drude (1852–1933), and it made him one of the founders of plant ecology. Drude was a remarkably talented field botanist. He could walk through a wood or across a meadow, identifying the plants as he went and noting local differences.

Drude wrote and coauthored several other important books. These included *Atlas der Pflanzenverbreitung* (Atlas of plant distribution), published in 1887, and *Handbuch der Pflanzengeographie* (Handbook of plant geography) in 1890. He coedited with Adolf Engler (1844–1930) *Die Vegetation der Erde* (The vegetation of the Earth), which appeared between 1896 and 1928.

Carl Georg Oscar Drude was born in Brunswick, Germany, on June 5, 1852. He studied natural history and chemistry at the Collegium Carolinum (now the Technical University) in Brunswick, moving to the University of Göttingen in 1871, where he worked as an assistant to August Grisebach (see "August Grisebach and Floral Provinces" on pages 122–124). Drude received his Ph.D. at Göttingen in 1873 and became an assistant to the plant collector Friedrich Gottlieb Theophil Bartling (1798–1875), who was in charge of the university's herbarium. From 1876 until 1879, Drude was a lecturer at the university.

In 1879, Drude moved to Dresden to take up an appointment as professor of botany at the polytechnic. In 1890 the polytechnic became a new technical high school, and Drude was in a strong position to influence the organization of scientific research and teaching. It was also in 1890 that Drude became director of the Dresden Botanical Gardens. He served as rector of Dresden Technical High School twice (1906–07 and 1918–19), and in 1920 he was made professor emeritus of botany. Drude died at Bühlau, near Dresden, on February 1, 1933.

EUGEN WARMING AND THE PRINCIPLES OF PLANT ECOLOGY

The first textbook on plant ecology was published in 1895 in Copenhagen, Denmark, entitled *Plantesamfund: Grundtræk af den økologiske Plantegeografi.* It was translated into German in 1896 and again in 1902, into Polish in 1900, and into Russian in 1901 and again in 1903. It first appeared in English in 1909, published by the Clarendon Press, Oxford, England, with the title *Oecology of Plants: An Introduction to the Study of Plant-Communities.* The book was based on the lectures that provided the framework for the world's first university course in plant ecology. The teacher and author of the textbook was Professor Eugenius (usually called Eugen) Warming (1841–1924) of the University of Copenhagen, a gifted and popular teacher whose ideas quickly became influential, despite the fact that he wrote in Danish, a language few foreign scientists could read.

Warming's course and his book introduced all of the world's major *biomes*—the largest biological units recognized, covering large areas and coinciding approximately with climatic regions. While describing the biomes, Warming explained how they developed. He showed that biological communities tended to solve particular environmental problems in similar ways, so that similar environments usually produced similar communities. These communities might consist of entirely different and only distantly related species, but they survived despite their differences because they had adapted to the conditions in which they lived. Cacti, for instance, are adapted to arid conditions, and they occur naturally only in the Americas. Their Old World equivalents, found in African deserts, are euphorbias, some of which are so similar to cacti that the two can be difficult to tell apart. Sidewinding, a method of locomotion that allows a snake to move across loose sand, is often associated with the American sidewinder (*Crotalus cerastes*), which is a species of rattlesnake. But at least three species of snakes that inhabit Old World deserts also employ sidewinding. Warming never accepted Darwin's theory of evolution by natural selection, however. A strong believer in adaptation, he believed that new species emerged through offspring inheriting characteristics their parents had acquired during their lifetime by adapting to local conditions. This was the theory, now rejected by biologists, that Lamarck had proposed (see "Erasmus Darwin and *The Botanic Garden*" on pages 175–177).

Johannes Eugenius Bülow Warming was born on November 3, 1841, on the Danish island of Mandø, where his father, Jens Warming (1797–1844), was the minister. His mother was Anna Marie von Bülow af Plüskow (1801–63). Following his father's death, Warming and his mother moved to Vejle, in Jutland. Warming attended the cathedral school in the town of Ribe, and in 1859 he began to study natural history at the University of Copenhagen. He interrupted his studies from 1863 to 1866 to work as a secretary to the Danish paleontologist Peter Wilhelm Lund (1801–80) in the tropical grasslands of Brazil. Warming obtained his doctorate in 1871 from the University of Copenhagen. He was a lecturer in botany at the university from 1873 to 1882, when he was appointed professor of botany at the Royal Institution, Stockholm, Sweden. He held this post until 1885. From 1886 to 1911, Warming held the position of professor of botany and director of the botanical gardens at the University of Copenhagen, but he was absent from Denmark from 1890 to 1892, engaged in fieldwork in Venezuela and the West Indies; in earlier years fieldwork took him to Greenland, Norway, and the Faroe Islands. He was rector of the university from 1907 to 1908. After his retirement, his successor as professor of botany was Christen Raunkiær (see "Christen Raunkiær and the Way Plants Grow" on pages 182–184).

Warming received many honors. He was a member of the Royal Danish Academy of Sciences and Letters from 1878 and an honorary fellow of the Royal Society of London. Warming married Johanne Margrethe Jespersen (1850–1922) in 1871. They had eight children. Eugenius Warming died in Copenhagen on April 2, 1924.

ARTHUR TANSLEY AND THE PLANTS OF BRITAIN

Ecology is now accepted as a scientific discipline in its own right, and there are professional ecologists. This is due in no small measure to the influence of another gifted teacher and writer, the English ecologist Arthur G. Tansley (1871–1955). Tansley always insisted on strict definitions, rigorous use of language, and learning about natural communities by studying the flow of energy and nutrients through them. He strongly opposed loose comparisons between plant and animal communities and human societies.

Tansley advised the student of ecology to begin by studying plant ecology. The fact that plants provide the basis for all terrestrial life makes it impossible to study animal ecology without referring to plants and, therefore, to plant ecology. Animals are relevant, of course, through their effects on plants, and so are humans. In his textbook *Practical Plant Ecology*, published in 1923 and in a revised and enlarged edition as *Introduction to Plant Ecology* in 1946, Tansley wrote the following:

> Thus anything like a *complete* study of the ecology of a plant community necessarily includes a study of the animals living in or feeding upon it. The influence of man upon plant communities is of first importance in all but the most uninhabited and the most sparsely inhabited regions of the earth. . . . we can never afford to lose sight of past and present human activities in their effects on the vegetation of countries which have long been inhabited and densely populated, like those of Western and Central Europe.

Tansley introduced the term *ecosystem* in "The Use and Abuse of Vegetational Concepts and Terms," an article published in the journal *Ecology* in 1935, but he did not coin the word. It was first used in 1930 by the British botanist Arthur Roy Clapham (1904–90) in response to an inquiry by Tansley, who had asked him to think of a word to describe the physical, chemical, and biological components of an environment when these are considered together. Tansley defined the term in *Introduction to Plant Ecology,* where he used biome in a different sense from the modern one, using it to describe "the whole complex of organisms—both animal and plants—naturally living together as a sociological unit," and continued:

> A wider conception still is to include with the biome all the physical and chemical factors of the biome's environment or habitat—those factors which we have considered under the headings of climate and soil—as parts of one physical *system,* which we may call an *ecosystem,* because it is based on the *oikos* or home of a particular biome.

Arthur George Tansley was born in London on August 15, 1871. He attended Highgate School from 1886 until early 1889, when he entered University College London to study science. He studied botany at Trinity College, University of Cambridge, graduating in 1894. After graduating from Cambridge, he returned to University College as a demonstrator in botany and assistant to his former teacher, Francis Oliver (1864–1951).

Tansley spent the years 1900 and 1901 studying the plants of Egypt, Sri Lanka, and the Malay Peninsula. Following his return to Britain, in 1902 he founded *The New Phytologist—phytology* is the study of plants. He was editor of this journal from its first issue until 1931. In 1907 Tansley became a lecturer in botany at Cambridge, where he remained until 1923. He had grown interested in psychology, and in 1923 he resigned from Cambridge and spent a year in Vienna studying under Sigmund Freud. In 1927 Tansley was appointed Sherardian Professor of Botany at the University of Oxford, where he remained until he retired in 1939.

Tansley was the first president of the British Ecological Society in 1913 and editor of the society's *Journal of Ecology* from its first edition in 1916 until 1938. He was elected a fellow of the Royal Society in 1915. In 1941 Tansley was awarded the gold medal of the Linnean Society. He was knighted in 1950. Sir Arthur Tansley died at Grantchester, near Cambridge, on November 25, 1955.

Biodiversity and Plant Conservation

In June 1992 the United Nations held a Conference on Environment and Development (UNCED) in Rio de Janeiro, Brazil. Popularly known as the Earth Summit and the Rio Summit, 178 heads of government attended UNCED. The conference agreed on a number of treaties, one of which was a Convention on Protecting Species and Habitats, which was soon renamed the Convention on Biological Diversity. The title is often shortened to the Biodiversity Convention. Its objectives are set out, in the rather stilted language employed in all intergovernmental agreements, in the preamble to the convention:

> The objectives of this Convention, to be pursued in accordance with its relevant provisions, are the conservation of biological diversity, the sustainable use of its components and the fair and equitable sharing of the benefits arising out of the utilization of genetic resources, including by appropriate access to genetic resources and by appropriate transfer of relevant technologies, taking into account all rights over those resources and to technologies, and by appropriate funding.

The convention was opened for the signature of government leaders who were present at UNCED on June 5, 1992. By the end of the summit, 150 heads of government had signed, and it came into force on December 29, 1993. By April 2009, 191 nations had joined.

This chapter discusses *biodiversity*, which is a contraction of biological diversity, with particular reference to plants. It begins by attempting to define the term and to explain why it is thought to be important. It continues by describing the historical effect on natural plant communities of the development of agriculture. The loss of natural vegetation to agricultural expansion and, more visibly though less significantly, to urban development generated a reaction that gave rise to the conservation movement. Areas were afforded legal protection from inappropriate exploitation. The chapter describes the origin and development of national parks and other protected areas in the United States and in other countries throughout the world. Finally, the chapter outlines the situation of the tropical forests.

WHAT IS BIODIVERSITY?

There are certain words that everyone uses but that are exceedingly difficult to define precisely. *Biodiversity* is one such word: The term is a contraction of biological diversity, which seems simple enough. Obviously, it means the variety of living organisms that inhabit our planet.

Unfortunately, the apparently simple definition leads to difficulties. If it refers to species, there are more than 24 different definitions for the word "species," so anyone proposing a scheme to save species from extinction must first decide which definition to use, because an organism that qualifies as a species under one definition may not do so under another. Perhaps, though, biodiversity refers to communities of organisms, but it would clearly be impossible to preserve every ecological community. The concept of evolution by natural selection is founded on the observation that there is considerable variation within any population of a species (however defined). Every individual is different. So perhaps efforts should be directed toward preserving every individual? But this is absurd, because it would mean abolishing death. In the end, the term has to remain vague, and we must accept that a term, describing an idea, may be useful even though it cannot be defined rigorously.

Plants are woven into the culture of almost all human societies. William Shakespeare is believed to have been a keen gardener and certainly he mentioned plants frequently in his plays and sonnets. These references would have been familiar to his audiences, as would the symbolism attached to the different flowers and herbs. New York's Central Park contains a Shakespeare Garden, displaying only plants that are mentioned in Shakespeare's works. There are also Shakespeare gardens in Brooklyn, Cleveland, Chicago, San Francisco, Johannesburg, Vienna, and many other cities. In the Middle Ages, the War of Roses, fought for the English throne, was between the houses of Lancaster—the red rose—and York—the white rose. The Tudors symbolized their reign with the Tudor rose, which is partly white and partly red. The flag of Lebanon features a cedar tree—the cedar of Lebanon (*Cedrus libani*)—and every nation and most states, provinces, counties, and regions within nations have their own symbolic plants.

The loss of a plant with such cultural, historical, or heraldic associations would be tragic. It would be felt as an assault on the culture itself. The protection of local or national heritage is important. Plants do not grow in isolation, however. They live in communities on which they depend, so protecting them requires preserving their communities—in other words, the biodiversity of the area where they are found.

Wild plants are also important for other reasons. Crop plants are descended from wild ancestors, and from time to time plant breeders and geneticists need to revisit those ancestors in search of particular qualities the crop has come to need. This is most commonly resistance to disease, but it might be tolerance of drought, or cold, or of soils contaminated with salt.

It is also possible that in years to come new plants will be domesticated in order to provide food or fiber. At present the world relies on a range of crop plants that may need to be augmented to meet the needs of future generations. The United Nations Food and Agriculture Organization (FAO) monitors the use of wild plants, especially those with the potential for domestication. Should the communities containing these plants disappear, this resource for the future would be lost.

Centuries ago physicians were taught botany because most medicines were derived from herbs cultivated for the purpose. Those medicines are now manufactured chemically on an industrial scale, but pharmaceutical research continues to explore wild plants in search of substances with therapeutic potential. If the wild plants disappear, that opportunity will also be lost.

There are clear aesthetic, cultural, agricultural, and pharmaceutical reasons for preserving natural plant communities. It may be impossible to define biodiversity rigorously, but there can be no doubt of its importance.

THE ADVANCE OF AGRICULTURE AND THE RETREAT OF WILDERNESS

People began to cultivate crop plants about 11,500 years ago in Southwest Asia and more recently in every other part of the populated world (see "The Origins of Agriculture" on pages 127–130). At first, the early crops faced severe competition from wild plants, but in time the farmers overcame them, at least partially. It was not only their chosen plant species that the farmers were domesticating; they were also domesticating their soils. In those parts of the world where the land has been farmed for thousands of years little evidence remains about the effect of domesticating the soil, but the history of arable farming in the United States offers a recent, and very dramatic, example of soil domestication.

Originally, the North American Great Plains were covered by temperate grassland—the prairies. Prairie grasses were the predominant plants, some of them growing as tussocks. The tussocks slowed the wind close to ground level and the mats of grass roots, some extending to a considerable depth, bound the soil particles together. The plants secured the surface soil from erosion in a windy and fairly dry climate.

Farmers began migrating onto the plains in large numbers around the middle of the 19th century, many drawn there by the Homestead Act of 1862, which offered a free plot of land to anyone willing to live on it and farm it. The farmers burnt off the prairie grasses and plowed the land. Plowing was hard work, because the soil naturally formed clods that baked rock-hard in the hot summer sunshine. These had to be broken up in order to produce a fine texture suitable for sowing with arable crops. The work became easier after about 1915, when tractors began to appear. Between 1910 and 1919 the area of arable farmland on the prairies more than doubled. Little by little the work grew less arduous and yields increased. The farmers were succeeding in "taming" the prairie soils.

Then, the economic depression of the 1930s forced farmers to intensify production in order to maintain their incomes, just as a series of years with above-average rainfall came to an end. A prolonged drought destroyed crops. In previous droughts the roots of the prairie grasses and the hard clods of earth had prevented erosion, but the fine-textured farm soils had no such protection. The soil simply blew away.

This is an extreme example, but everywhere in the world, farming the land involves clearing the natural vegetation and progressively altering the character of the soil. Over most of lowland Europe, for instance, the temperate forest was the original natural vegetation. Except in the far north, where the climate is too cold for arable farming, farmers cleared almost all of that forest centuries ago. The landscapes of Europe are beautiful, but they are the product of agriculture.

The total land area of the world is approximately 50.40 million square miles (130.58 million km^2). Of that total, almost 44 percent is farmed either with arable crops, perennial crops, or permanent pasture. Today, most people live in urban areas, and the expansion into the countryside of housing, commercial development, and roads is highly visible. Understandably, people fear that this poses a serious threat to biodiversity. Globally, however, urban areas occupy a mere 1.5 percent of the land area. It is not urban expansion that threatens biodiversity, but the expansion of agriculture.

During the 21st century, the predicted increase in the size of the human population means that food production must increase substantially. Every available technology will have to be applied if the increased output is to be achieved without destroying even more areas of natural habitat and reducing biodiversity. From the point of view of preserving biodiversity, the greenest type of farming is that which produces the highest yields from the smallest area of land while using agricultural chemicals efficiently to minimize both cost and pollution.

NATIONAL PARKS AND NATURE RESERVES

In 1830 the American painter and author George Catlin (1796–1872) set off on a journey up the Mississippi River into Native American lands at the start of a diplomatic mission led by General William

Clark (1770–1838). By 1836 Catlin had visited 50 tribes, and in the following two years he visited 18 more on a journey up the Missouri River. As he went, Catlin painted some of the people he met, and when his travels ended he exhibited his paintings in the United States and later across Europe.

Catlin's primary aim was to earn a living from his paintings, at which he was not very successful, but he also had another purpose. He had realized that the westward expansion of agriculture, which would soon be followed by industry and urbanization, threatened the way of life of the peoples he had grown to admire. He believed they should be protected, and to this end he proposed the creation of a national park that would allow people and wildlife to flourish undisturbed.

A version of Catlin's idea aroused interest, and in 1864 Congress presented an area of land to the state of California to be preserved as a state park. The state park was then enlarged and it became a national park on October 20, 1890. On March 1, 1872, an act of Congress designated land near the headwaters of the Yellowstone River, as a public park, making Yellowstone the world's first national park. Yosemite and Sequoia national parks were designated in 1890 and others followed. The first forest reserve was established in 1891, and the first national wildlife refuge in 1903. The United States now has 6,770 areas of federal protected lands with a combined area of more than 1 million square miles (26 million km^2). The Royal National Park, established in 1879, was the first in Australia, and in 1885 the Rocky Mountains National Park became Canada's first national park. Sweden was the first European country to designate national parks, in 1909. Today, worldwide, there are more than 6,500 national parks—the largest of which is Northeast Greenland National Park, which covers 375,000 square miles (972,000 km^2). There are now national parks on every continent, and in Africa there are also game reserves set aside for that continent's large mammals.

National parks protect the landscapes, wildlife, and natural features of relatively large areas. Nature reserves protect particular habitats and the species they sustain. Many are small in area and they are widely scattered, but their combined area is considerable. Germany, for instance, has 5,314 nature reserves with a total area of 2,642 square miles (6,845 km^2).

Areas of worldwide importance for biodiversity may be designated as biosphere reserves under the Program on Man and the Biosphere administered by UNESCO (the United Nations Educational, Scientific and Cultural Organization). Biosphere reserves may be on land or at sea; the aim of identifying them is to promote and demonstrate a balanced relationship between humans and the biosphere, and under the official regulations governing the scheme, in order to qualify an area must "encompass a mosaic of ecological systems." There are 531 biosphere reserves in 105 countries. The United States has 47 and Canada has 15. The United Kingdom has 10.

There is now a worldwide network of land and sea areas that are afforded protection from exploitation that might damage their wildlife. This helps greatly in preserving species and habitats—in protecting biodiversity.

SAVING THE TROPICAL FORESTS

Tropical forests cover about 6 percent of the Earth's land area. It is difficult to be precise about the area, but in 2009 tropical forests of all types probably occupied approximately 7 million square miles (18 million km^2). As well as rain forests, the Tropics support seasonal forest, dry forest, and mangrove forest, and all of these vary in structure and composition from one region to another. Tropical forests occur in at least 90 countries. The largest area is in Latin America and the Caribbean, which has about 52 percent of the total tropical forest, followed by Africa with 30 percent, and Asia and the Pacific Islands with 18 percent.

In some places, tropical forests occupy land that is in demand for food production, and in most regions where this is the case small-scale farmers and agricultural companies have converted substantial areas of forest to farmland. The forests also contain trees that yield high-value hardwood timber, and in some areas illegal logging is clearing some forest and damaging more by opening tracks to gain access to selected trees. Roads driven through the forest encourage agricultural expansion by allowing farmers to penetrate the forest and clear land to the sides of the roads. Mangrove forests in Asia are being damaged by the expansion of shrimp farming in coastal waters.

News stories about areas of forest clearance, often accompanied by photographs showing forests devastated by machinery, have led many people to fear that the tropical forests are being subjected to such a ferocious attack that within a few decades they may disappear altogether. It is certain that the area of tropical forests has decreased substantially in the past, but it may be that the decline has been slowed and perhaps halted. There was no decrease, and possibly a small increase, in the area of tropical rain forest between 1983 and 2000, measured in 63 countries; another study found no decrease in area since the early 1970s in the same 63 countries. The improvement was due partly to the success of programs to protect tropical rain forest and partly to natural regrowth in areas that had been cleared. The steady increase in the atmospheric concentration of carbon dioxide has also increased the rate of growth in tropical forests. The FAO predicts a 10 percent increase in the total area of tropical forest by 2050.

There is considerable uncertainty about the data, however. The FAO takes great care in compiling its assessments, but it has to base them on data submitted by national governments, which may contain errors, augmented with satellite imagery. The overall picture also conceals large local variations, and rapid deforestation is continuing in some countries. Some conservationists question the definitions of forest; certain areas the FAO classes as forest are more like savannah grassland with scattered trees.

Their warm climate and long growing season allows tropical forests, and especially tropical rain forests, to sustain a wide range of species that can survive nowhere else. If the forests are cleared many of those species will disappear. If entire forests were cleared this would obviously be true, but it is not necessarily true of clearance on a smaller scale. The extent of the threat is based on counting the number of species in a number of measured areas and extrapolating those numbers to very much larger areas. This method tends to overestimate biodiversity, because some parts of a forest support more species than others and many species occur over wide areas, so their contribution to biodiversity is counted several times. In areas where the species were counted before the forest was cleared, scientists have found that most of them survived in the adjacent forest. In Southeast Asia, where plantations with one or just a few tree species have replaced large areas of natural forest, plants and animals have survived in the remaining forest and the number of endangered species has not increased.

There is clear historical evidence for the recovery of tropical rain forests. At one time, extensive areas were farmed along river valleys in the Amazon Basin. Estimates of the total farmed area range from 0.1 percent to 0.3 percent of the entire basin, which is 2,300–3,000 square miles (6,000–8,000 km²), with villages every few miles. Archaeologists believe farming began 1,000 to 2,000 years ago and disappeared soon after the arrival of Europeans. The forests then reclaimed the land, and the area is now ecologically indistinguishable from the areas that were never farmed.

Deforestation in the Tropics may be slowing overall, but it continues in some regions and very large areas have been lost already. The threat that forest clearance poses for biodiversity may have been exaggerated, but that does not mean there is no threat at all. Efforts to help the governments of tropical countries to protect their forests deserve support.

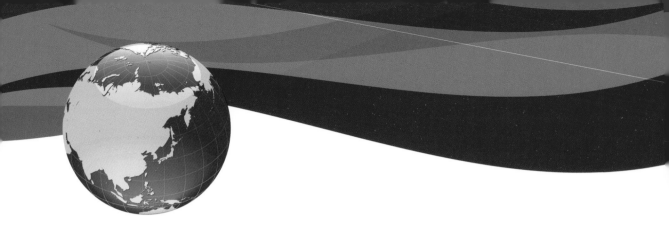

Conclusion

As biologists have learned more about the physiology of plants, the biochemical reactions involved in their growth, maintenance, and reproduction, and in their genetic composition, it has proved necessary to rename the science of botany. The study of plants is no longer a single discipline, but a group of related disciplines that are known collectively as the plant sciences. Scholars have been studying plants for thousands of years, and they have found, as researchers in every other area of science have found, that the more they learn the more unanswered questions they discover. Science is a never-ending process.

This short book has traced the history of plant science and plant use. It has described the medicinal uses of plants and how their cultivation led to the invention of gardens. It has outlined the development of written descriptions and catalogs of plants, which revealed a need for a system of plant classification that all naturalists could accept and use. As explorers visited distant lands they discovered plants that were unknown back home, and plant hunters traveled the world in search of species that could be cultivated for ornament or use. The book has told of the origins of plant cultivation, the unraveling of plant evolution, and the emergence of plant ecology.

The skills of plant scientists are needed now as never before. Unless food production increases rapidly, the predicted rise in the world population means there could be serious shortages by the middle of the 21st century. Plant scientists will need to develop new crop varieties that yield more, make more economical use of water, and are more nutritious than those being grown today. They will have to employ traditional breeding techniques and genetic modification—not least to minimize dependence on costly agricultural chemicals.

Plant scientists also have another task, to identify, classify, and record the plant species that are growing in areas threatened by agricultural, forestry, or urban expansion. Without this knowledge it will be impossible to assess the extent of biodiversity and the risks to it. Among the previously unknown plants, there may be some that have the potential to become important sources of food, fiber, or drugs.

There is no shortage of urgent tasks facing plant scientists. Perhaps, having read this book, you would like to become one of them and contribute to the important and exciting discoveries that will certainly emerge in years to come.

GLOSSARY

acotyledon   a plant that has no **cotyledons** in its embryo.

alga   a simple plantlike organism that performs **photosynthesis** but is not differentiated into roots, stems, and leaves.

anatomy   the study of the structure of living organisms.

annual   a plant that completes its life cycle, from germination to releasing seeds, in a single year.

anther   the clublike male reproductive of a **flower**.

antihelminthic   a drug that expels parasitic intestinal worms, either by killing or stunning them.

apothecary   a chemist who makes (and usually sells) medicinal drugs.

arboretum   a collection of living tree species.

archaeobotanist   a botanist who studies plant remains recovered from archaeological sites.

awn   a long bristle that protrudes from a cereal grain.

biennial   a plant that completes its life cycle in two years.

binomial system   the use of two names to identify a species, the first name being that of the genus and the second of the species; by convention both names are printed in italic type and the generic name is written with an initial capital letter.

biodiversity   a contraction of biological diversity.

biogeography   the study of the geographic distribution of plants and animals.

biome   one of the largest biological units recognized, covering a large area and coinciding approximately with a climatic region.

boreal   pertaining to the north, from the name of Boreas, the god of the north wind in Greek mythology.

boreal forest   the coniferous forest of subarctic climates that forms a belt across northern North America and Eurasia.

bract   a leaflike structure growing from the point where a **peduncle** joins the main stem of a plant.

brazilwood   the tree *Caesalpinia echinata,* native to tropical South America, that yields a red dye and from which the country of Brazil acquired its name.

bulb   an underground storage organ (**perennating bud**) consisting of a short, fat stem with roots at its base above which there are fleshy leaves surrounded by protective scales.

calyx   in a **flower**, the set of **sepals** that enclose and protect the flower bud.

carpel   the female reproductive organ in a flower, comprising the **stigma**, **style**, and **ovary**.

carpellate   of a flower, possessing only a **carpel**.

cell theory   the fundamental tenet of biology, which states that all living organisms either consist of a single cell or are made up of cells and that they grow and reproduce by the division of cells.

chamaephyte   a plant in which the **perennating buds** occur very close to the ground.

chlorophyll   the green pigment in **chloroplasts** that absorbs light (principally at the blue and red wavelengths), initiating the series of reactions comprising **photosynthesis**.

chloroplast   the body within plant cells that contains **chlorophyll** and is the site of **photosynthesis**.

cline   a gradual physiological or behavioral change in members of a species across the geographic range of that species.

corolla   the structure formed by the petals of a **flower**.

cotyledon   seed leaf; the first leaf to appear from a germinating seed.

cryptogams   plants that reproduce by spores rather than seeds, including algae, lichens, mosses, and ferns.

cryptophyte   a plant in which the **perennating buds** are below the ground or water surface.

cyanobacterium   a bacterium that performs **photosynthesis**.

cytoplasm   all the material enclosed by a cell membrane, except for the nucleus.

cytoplasmic streaming   *see* **streaming**.

dicotyledon   a plant that produces two or more **cotyledons** and leaves with a network of veins.

dioecious   having separate male and female plants.

disjunct distribution   the occurrence of closely related species in scattered locations separated by substantial barriers to migration such as oceans.

ecology   the scientific study of the relationships among living organisms and between living organisms and their living and nonliving environment.

ecosystem   a discrete unit comprising living and nonliving components that interact to form a stable system.

endosperm   a store of food contained in a seed that sustains the young plant until it is able to obtain food for itself.

ephemeral plant (ephemerophyte)   a plant that completes its life cycle from germination to the release of seeds during a brief period of favorable conditions.

ephemerophyte   *see* **ephemeral plant**.

epiphyte   a plant that grows on the surface of another plant, using that plant only for support and not as a source of nutrients.

filament   the stalklike structure that supports the **anther** in a male **flower**.

flora   all the plants, useful or not, growing in a specified area, or a description of them.

floral province   a group of plants covering a large geographic area, all of which are adapted to the climate of that area.

flower   the reproductive structure of a flowering plant.

formation   an assemblage of plants determined by climate.

Fungi   a taxonomic kingdom comprising organisms that obtain nutrients by absorbing them from their surroundings; they do not perform **photosynthesis**.

fungus   an organism belonging to the kingdom **Fungi**.

geophyte   a **cryptophyte** in which the **perennating buds** are below ground level.

helophyte   a **cryptophyte** in which the **perennating buds** are below the surface of a marsh.

hemicryptophyte   a plant in which the **perennating buds** are at ground level.

herbarium   a collection of preserved plant specimens or the building in which such a collection is housed.

holotype (type specimen)   an individual plant, usually preserved as a **herbarium** specimen but sometimes as a botanical illustration, to which a taxonomic name refers. Other plants are classified in the same species only if they match the holotype.

homology   the existence in two species of organs or structures that appear different but that are descended from a common ancestor.

hurdle   a temporary fence, often made from woven willow, hazel, or ash, that can be moved as required.

hydrophyte a **cryptophyte** in which the **perennating buds** are below the surface of water.

hygrophilous growing in or preferring moist habitats.

imperfect flower a **flower** lacking either petals or **sepals**.

inflorescence a flowering structure consisting of more than a single **flower**.

isoline a line joining places with similar values for a specified variable.

isotherm a line joining points on the Earth's surface or at the same elevation with respect to the surface, where the temperature is the same.

latex a milky fluid produced by some herbs and trees that may carry nutrients and may also help the plant to heal wounds.

lectotype a member of a **syntype** that is chosen, after the publication of the original description, as a **holotype**.

liana *see* **liane**.

liane (liana) a free-hanging climbing plant.

lichen a composite organism consisting of a **fungus** and an **alga** or **cyanobacterium**.

ligneous woody.

microclimate the atmospheric conditions found in a small local area.

monocotyledon a plant that produces a single **cotyledon** and leaves with parallel veins.

monoecious bearing male and female flowers on the same plant.

mosaic virus a virus that causes an infected plant to produce petals or leaves with specks or patches of color.

mycology the scientific study of **fungi**.

order bed a plant bed in which all the plants belong in a particular category, such as order, family, or genus; most botanical gardens arrange their plants in order beds.

ovary in a female **flower**, the structure holding the eggs (ovules).

ovule *see* **ovary**.

paddy a shallow pond in which rice is grown.

paddy rice rice that has been grown in a **paddy**.

paleobotany the scientific study of plant fossils and other traces and remains in order to reconstruct past environments and the evolutionary history of plants.

paleoclimatologist a scientist who studies the ancient history of climate.

palynology   the study of pollen grains and the spores of nonflowering plants.

panicle   a compound, many-branched **inflorescence**.

paratype   a plant specimen, other than a **holotype**, that an author uses in the original description of a new species.

parenchyma   plant tissue consisting of unspecialized cells with air spaces between them.

parterre   a level garden area with flower beds.

peduncle   the stalk holding a **flower**.

perfect flower   a **flower** possessing both **stamens** and **carpels**.

perennating bud (perennating organ)   the plant structure with which a **perennial** plant survives periods of adverse conditions.

perennating organ   *see* **perennating bud**.

perennial   a plant that lives for more than two years and flowers annually after an initial period during which it may not flower at all.

phanerophyte   a plant in which the **perennating buds** are above ground on shoots exposed to the air.

photosynthesis   the series of chemical reactions, driven by the energy of sunlight, by which green plants and certain bacteria synthesize carbohydrates out of carbon dioxide obtained from the air and water absorbed from the soil.

physiology   the scientific study of the way plants and animals work.

phytogeography   the study of the geographical distribution of plants.

phytology   the study of plants.

phytopathology   the study of plant diseases.

phytosociology   the classification of plant communities according to the characteristics and relationships of and among the plants within them.

plumule   the tip of a shoot or bud on an **embryo**.

pollen   the mass of grains produced within the **anther** of a **flower**, each of which contains two sperm nuclei and one nucleus that gives rise to the **pollen tube**.

pollen tube   the tube that grows from a **pollen** grain and enters the **ovary**, and down which the two sperm nuclei move from the pollen grain.

protoplasm   the colorless, translucent contents of a living cell, including the cell membrane but excluding the large cell **vacuoles**, ingested material, and secretions. It is divided into the protoplasm inside the nucleus and is known as **cytoplasm** in the rest of the cell.

protoplasmic streaming   *see* **streaming**.

quadrat   the basic ecological sampling area for studying plant communities, commonly 1m^2 for grassland but larger for woodland. The plot is marked on the ground, and the researcher counts every plant of every species inside the boundary.

rachis   the axis of a cereal ear or compound leaf.

radicle   the rudimentary root that grows from an **embryo**.

Raunkiær's law   a law stating that the plant species growing in a specified area will be either very common or very rare.

recalcitrant seed   a seed that cannot survive being dried and that germinates rapidly after the plant releases it. Many plants of the tropical rain forest produce recalcitrant seeds. Such plants cannot be conserved in **seed banks**, but only by being grown.

receptacle   the enlarged end of the **peduncle** at the base of a **flower**.

relative humidity   the amount of water vapor present in the air as a percentage of the amount required to saturate the air at that temperature.

relevé   the basic field recording unit in **phytosociology**; it should be an area with uniform vegetation, topographical relief, and type of soil, and the record includes information about the soil and other environmental features.

rhizome   a horizontal, creeping, underground stem from which roots and shoots emerge at intervals and that serves as a **perennating bud**.

seed bank   a store in which plants are preserved as seeds.

sepal   in a **flower**, one of the leaflike structures that together form the **calyx**.

simple   a medicine derived from a single ingredient.

simplicia   plants with medicinal properties.

spontaneous generation   the idea, prevalent in the Middle Ages, that living animals could emerge from putrefying matter.

stamen   the male reproductive organ of a flower, comprising an **anther** and **filament**.

staminate   of a flower, possessing only **stamens**.

stigma   in a female **flower**, the structure at the top of the **style** to which pollen grains adhere.

stoma   *see* **stomata**.

stomata (sing. stoma)   the pores in leaf surfaces through which a plant exchanges gases and loses moisture through **transpiration**.

streaming (cytoplasmic streaming; protoplasmic streaming)   the movement of **protoplasm** within and between living cells, and along fibers composed of elongated cells joined end to end.

style   in a female **flower**, the structure linking the **stigma** and **ovary**.

syntype   a set of plant specimens to which no **holotype** has been assigned.

taxidermy   the craft of posing dead animals in lifelike attitudes for display.

taxonomy   biological classification.

terrarium (Wardian case)   a tightly sealed container for living plants and small land-dwelling animals.

therophyte   a plant that completes its life cycle rapidly during times when conditions are favorable, then dies down and survives as seed.

topiary   the art of clipping trees and shrubs to make sculptured shapes.

transpiration   the movement of water through a plant from the roots to the leaf surfaces where it is lost by evaporation through the **stomata**.

tuber   a swollen stem or root that serves as an underground storage organ (**perennating bud**).

type specimen   *see* **holotype**.

vacuole   a sac, enclosed by membranes, that is found inside a cell, where it serves as a storage organ.

vicariance   the occurrence of two closely related species at widely separated locations, so that each species is the geographic equivalent of the other.

Wardian case   *see* **terrarium**.

FURTHER RESOURCES

Allaby, Michael. *Droughts,* rev. ed. New York: Facts On File, 2003. Contains an account of the dust bowl drought of the 1930s.

Allaby, Robin G. "The rise of plant domestication: life in the slow lane." *Biologist* 55, no. 2 (May 2008). A fairly simple account of the evidence that cereal domestication took thousands of years.

Blunt, Wilfrid. *Linnaeus: The Compleat Naturalist.* London: Francis Lincoln, 2004. A full biography of Linnaeus.

Bowler, Peter J. *The Fontana History of the Environmental Sciences.* London: Fontana Press, 1992. A comprehensive history of the environmental sciences by the professor of history and philosophy of science at the Queen's University of Belfast.

Cooper, Alix. *Inventing the Indigenous: Local Knowledge and Natural History in Early Modern Europe.* Cambridge: Cambridge University Press, 2007. The story of the growth of interest in studying plants found locally.

Dash, Mike. *Tulipomania.* London: Victor Gollancz, 1999. A full account of the 17th-century Dutch tulip craze.

Fuller, Dorian Q., Ling Qin, et al. "The domestication process and domestication rate in rice: spikelet bases from the Lower Yangtze." *Science* 323 (March 20, 2009): 1607–1610. A scientific paper showing that the domestication of rice was a very slow process.

Heywood, V. H., R. K. Brummitt, A. Culham, and O. Seberg. *Flowering Plant Families of the World.* London: Royal Botanic Gardens, Kew, 2007. An illustrated work listing all the families of flowering plants, with details of distribution and uses.

Jardine, Lisa. *Ingenious Pursuits: Building the Scientific Revolution.* London: Little, Brown, 1999. An account of the development of modern science through the 17th and 18th centuries by a leading historian.

Judd, Walter S., Christopher S. Campbell, Elizabeth A. Kellogg, and Peter F. Stevens. *Plant Systematics: A Phylogenetic Approach.* Sunderland, Mass.: Sinauer Associates, 1999. A textbook on modern plant classification.

Ogilvie, Brian W. *The Science of Describing: Natural History in Renaissance Europe.* Chicago: Chicago University Press, 2006. An account of the development of the study of natural history in the 15th and 16th centuries.

Porter, Roy, and Mikulás Teich, eds. *The Scientific Revolution in National Context.* Cambridge: Cambridge University Press, 1992. A collection of essays on the development of science in France, German-speaking nations, Low Countries, Poland, Spain and Portugal, England, Bohemia, Sweden, and Scotland.

Tansley, A. G. *Practical Plant Ecology.* London: George Allen and Unwin, 1923. First edition of Tansley's classic textbook.

———. *Introduction to Plant Ecology.* London: George Allen and Unwin, 1946. Revised and enlarged second edition of *Practical Plant Ecology.*

WEB SITES

Brongniart, Adolphe-Théodore. *Prodrome d'une Histoire des Végétaux Fossiles.* Paris, 1828. Available online. URL: http://books.google.com/books?id=ivcTAAAAQAAJ&printsec=frontcover&dq=intitle:Histoire%2Bintitle:des%2Bintitle:Végétaux%2Bintitle:Fossiles&lr=&num=100&as_brr=1&as_pt=ALLTYPES#PPP5,M1. Accessed April 14, 2009. The full text of Brongniart's work in a facsimile edition (in French).

Chelsea Physic Garden. Homepage of the garden. Available online. URL: http://www.chelseaphysicgarden.co.uk/. Accessed February 25, 2009. Full description of the garden with details of opening times and events.

Culpeper, Nicholas. *The Complete Herbal.* Bibliomania. Available online. URL: http://www.bibliomania.com/2/1/66/113/frameset.html. Accessed February 20, 2009. The full text of Culpeper's work.

Darwin, Charles. *On the Origin of Species by Means of Natural Selection: Or The Preservation of Favored Races in the Struggle for Life.* Available online. URL: http://books.google.com/books?id=j-kRAAAAYAAJ&dq=Charles+Darwin&printsec=frontcover&source=an&hl=en&ei=U5rkSc_0OajslQfC6pDgDg&sa=X&oi=book_result&ct=result&resnum=8. Accessed April 14, 2009. The complete text of the 6th edition (1888) of Darwin's book.

Darwin, Erasmus. *The Botanic Garden.* Available online. URL: http://books.google.com/books?id=Tn8gAAAAMAAJ&dq=Erasmus+Darwin&printsec=frontcover&source=an&hl=en&ei=KJLwSdn2NdiDlAfZ453DDA&sa=X&oi=book_result&ct=result &resnum=5#PPA167,M1. Accessed April 23, 2009. The complete text of Darwin's long poem.

Denevan, William M. "The pristine myth: The landscape of the Americas in 1492." Available online. URL: http://jan.ucc.nau.edu/~alcoze/for398/class/pristinemyth.html. Accessed April 2, 2009. Article describing American agriculture prior to the arrival of Europeans.

Dodds, Justine. "Zoopharmacognosy: The medicinal use of plants by Chimpanzees in the wild." Available online. URL: http://www.behav.org/Student_essay/primates/ Dodds_2008_ChimpSelfMedication.pdf. Accessed February 11, 2009. An essay on self-medication by chimpanzees, illustrated with color photographs.

El-Abbadi, Moustafa. "The Library of Alexandria—Ancient and Modern." Hellenic Electronic Center and Professor Moustafa El-Abbadi, 1998. Available online. URL: http://www.greece.org/Alexandria/Library/index.htm. Accessed February 13, 2009. A full account of the great library.

Farlow, W. G. *Memoir of Asa Gray 1810–1888.* Available online. URL: books.nap.edu/html/biomems/agray.pdf. Accessed April 15, 2009. The full text of a paper read before the National Academy on April 17, 1889.

Food and Agriculture Organization of the United Nations. "Use and potential of wild plants in farm households." Available online. URL: http://www.fao.org/docrep/003/w8801e/w8801e00.htm. Accessed April 30, 2009. An FAO booklet on existing and potential uses of wild plants and on wild plants that might be domesticated.

Gilbert, L. A. "Banks, Sir Joseph (1743–1820)." *Australian Dictionary of Biography* (online edition). Canberra: Australian National University. Available online. URL: http://www.adb.online.anu.edu.au/biogs/A010051b.htm. Accessed March 6, 2009. A biography of Banks.

Hobson, Amanda. "Reginald Farrer of Clapham." North Craven Heritage Trust: *Journal 1992.* Available online. URL: http://www.northcravenheritage.org.uk/nchtjournal/Journals/1992/J92A13.html. Accessed March 18, 2009. An article about the life and work of Farrer.

Kohlhepp, Gerd. "Scientific findings of Alexander von Humboldt's expedition into the Spanish-American Tropics (1799–1804) from a geographical point of view." Rio de Janeiro: *Anais da Academia Brasiliera de Ciências,* 77, 2, June 2005. Available online. URL: http://www.scielo.br/scielo.php?pid=S0001-37652005000200010&script=sci_arttext. Accessed March 20, 2009. An article describing Humboldt's scientific achievement.

Lancelot "Capability" Brown. JevStar.com. Available online. URL: http://www.capabilitybrown.org.uk/. Last updated January 31, 2004. Accessed February 26, 2009. A site devoted to the life and work of this landscape gardener.

Moore, A. W. "Manx Worthies: Professor Edward Forbes." *The Manx Notebook,* vol. 3. Available online. URL: http://www.isle-of-man.com/manxnotebook/manxnb/v11p125.htm. Accessed March 31, 2009. Biography of Forbes.

Peithô's Web. "Life of Theophrastus," from *The Lives and Opinions of Eminent Philosophers* by Diogenes Laërtius, translated by C. D. Yonge. Available online. URL: http://classicpersuasion.org/pw/diogenes/dltheophrastus.htm. Accessed February 11, 2009. The full text of a translation, made in 1853, of Diogenes Laërtius's account of the life and works of Theophrastus.

"PlantExplorers.com: The Adventure Is Growing." Available online. URL: http://www.plantexplorers.com/index.html. Accessed March 6, 2009. A Web site devoted to the history of plant collecting.

Salisbury, Edward J. "Carl Johan Fredrik Skottsberg. 1880–1963." In *Biographical Memoirs of Fellows of the Royal Society* 10 (1964): 244–256. Available online. URL: http://rsbm.royalsocietypublishing.org/content/10/244.full.pdf+html. Accessed April 1, 2009. Biography of Skottsberg, a Foreign Member of the Royal Society.

Tansley, A. G. *Practical Plant Ecology.* Available online. URL: http://www.archive.org/details/PracticalPlantEcology. Accessed April 29, 2009. The full text of Tansley's 1923 book available to download.

Tooley, Oliver. "Frank Kingdon-Ward." Available online. URL: http://www.french4tots.co.uk/kingdon-ward/fkw-biog1.html. Accessed March 17, 2009. Biography of Kingdon-Ward by his grandson.

Tournefort, Joseph Pitton de. *Elements of botany or methodology by which to know about plants.* University of Tokyo. Available online. URL: http://edb.kulib.kyoto-u.ac.jp/exhibite/b06/b06cont.html. Accessed March 12, 2009. The full text and illustrations of Tournefort's 1694 work.

Turesson, Göte. "The Species and the Variety as Ecological Units," in *Hereditas.* Reproduced in *Early Classics in Biogeography, Distribution, and Diversity Studies.* Available online. URL: http://www.wku.edu/~smithch/biogeog/TURE1922.htm. Accessed April 16, 2009. The edited text of the 1922 essay in which Turesson introduced the term *ecotype.*

Wicke, Roger. "A World History of Herbology and Medical Herbalism: Oppressed Arts." Rocky Mountain Herbal Institute. Available online. URL: http://www.rmhiherbal.org/a/f.ahr1.hist.html. Updated January 19, 1998; accessed February 12, 2009. A brief history of the use of medicinal herbs.

INDEX